BASIC PHYSICS AND MEASUREMENT IN ANAESTHESIA

Basic Physics and Measurement in Anaesthesia

Second Edition

G. D. Parbrook
MD FFARCS

*Senior Lecturer, University of Glasgow
Department of Anaesthesia, The Royal Infirmary, Glasgow.*

P. D. Davis
BSc MInstP

*Principal Physicist
West of Scotland Health Boards' Department of
Clinical Physics and Bio-Engineering, Glasgow.*

E. O. Parbrook
BSc MB ChB

*Research Assistant
Department of Anaesthesia, The Royal Infirmary, Glasgow.*

APPLETON-CENTURY-CROFTS/Norwalk, Connecticut

0-8385-0621-6

Notice: The author(s) and publisher of this volume have taken care that the information and recommendations contained herein are accurate and compatible with the standards generally accepted at the time of publication.

First published
by William Heinemann Medical Books Ltd
23 Bedford Square, London WC1B 3HH

Distributed in the United States of America,
its territories and possessions and Canada
by Appleton-Century-Crofts,
25 Van Zant Street, East Norwalk,
Connecticut 06885

Copyright © G. D. Parbrook, P. D. Davis and E. O. Parbrook, 1986

All rights reserved. This book, or parts thereof, may not be used or reproduced in any manner without written permission. For information, address Appleton-Century-Crofts, 25 Van Zant Street, East Norwalk, Connecticut 06855.

86 87 88 89 90 / 10 9 8 7 6 5 4 3 2 1

Prentice-Hall Canada, Inc.

Library of Congress Catalog Number
ISBN: 0-8385-0621-6

Printed in Great Britain

Contents

Preface to First Edition

Our aim is to provide sufficient understanding of physics and its clinical application to allow the practising anaesthetist to give safe and reliable anaesthesia. In addition, this book should help to bridge the gap between the level of teaching in school and in undergraduate curricula and that in the more advanced textbooks on clinical physics. It is not intended to provide a rigorous treatment of physics or to act as a detailed reference text: consequently, references have been omitted. The use of mathematics, too, has been strictly limited and to render the subject more interesting to anaesthetists clinical examples are sometimes provided in detail even though this may involve digression into the realms of physiology and clinical anaesthesia.

Although written with the needs of the anaesthetist in mind, we hope the book may also be of interest to technical and senior nursing staff working in the operating theatre and intensive care areas.

Measurement is an integral part of physics and is included in the book because monitoring equipment plays such an important part in clinical anaesthesia. Nevertheless, all aspects of clinical measurement are not covered and details of physiological, biochemical and haematological measurement are not provided.

Previous experience in preparing an audiotape-slide series on *Clinical Physics and Measurement for Anaesthetists* has been helpful in the preparation of this book, but the content of the work is not identical with that of the audiovisual series and is so designed that it may be used either on its own or to complement the series.

The SI system of units is used throughout, and a resumé of the definitions of these units of measurement is given in the appendix. The exponent is used instead of the solidus because this is the more common form.

We are grateful to colleagues for their help and in particular to Dr W. G. Anderson for assistance with the chapter on Oxygen Measurement and to Dr R. Hughes for advice on the chapters on Blood Pressure Measurement and on Measurement of pH and CO_2. We are also grateful to Mr I. B. Monk for help with much of the original material.

We should also like to thank Mr B. D. Cameron, our graphic designer, for his expert production of the diagrams for this book.

1982

G. D. Parbrook
P. D. Davis
E. O. Parbrook

Preface to Second Edition

The steady expansion of the variety and complexity of the instruments used in the operating theatre, recovery and intensive care areas has led to an expansion in the size of the second edition. Apparatus now described includes the transcutaneous oxygen monitor, lasers, the oxygen concentrator, infusion controllers and pumps and the Emma and ultra-violet halothane analysers. Safety regulations and precautions have become more exacting and the chapter on Electrical Safety has been extensively rewritten.

In this edition, consideration of gas supplies and suction is no longer restricted to the supply to the anaesthetic machine. An extra chapter has been added on Breathing and Scavenging Systems to bridge the gap between the machine and the patient.

In keeping with the aims of the book, more examples of the clinical application of physical principles are included; for example, calculations are now provided for typical pressures generated in syringes and for the heat changes with shivering or following blood transfusion.

Diagrams are regarded as a vital aspect of teaching, so 35 new figures are provided in this edition and improvements made to many of those which appeared in the first edition.

The authors would like to thank all those who made constructive criticisms of the first edition. Their comments have been taken into consideration in the preparation of this edition.

1985

G. D. Parbrook
P. D. Davis
E. O. Parbrook

Audiovisual Series

The authors, with the help of colleagues, have also produced a series of audiotape-slide programmes entitled *Clinical Physics and Measurement*, which is published by Oxford Educational Resources Ltd.

The publishers, William Heinemann Medical Books Ltd, would like to express their appreciation to Oxford Educational Resources Ltd for permission to use many of the illustrations from the audiovisual series as a basis for those appearing in this book.

The content of the book and the audiovisual series are not identical but are worded to suit the medium concerned and the book or the series may be used either alone or to complement each other. Each audiovisual programme consists of 35 mm colour slides, an audio cassette, summary and questionnaire, and details may be obtained from William Heinemann Medical Books Ltd or direct from:

Oxford Educational Resources Ltd,
197 Botley Road,
Oxford OX2 0HE.

Telephone: 0865 726625
Telex: 837416—Oxstat G

CHAPTER 1

Pressure

INTRODUCTION, FORCE

As the anaesthetist makes pressure measurements both in patients and on the anaesthetic machine, an understanding of this topic is essential. Before concentrating on pressure, however, the definition of force must be considered. Force is that which changes or tends to change the state of rest or motion of an object. In the SI system, force is measured in newtons (N), a newton being the force that will give a mass of 1 kilogram an acceleration of 1 metre per second per second.

$$N = kg\ m\ s^{-2}$$

The force of gravity acting on any object will give the object an acceleration of $9.81\ m\ s^{-2}$. Therefore, the force of gravity on a mass of 1 kilogram must be 9.81 N. This force is known as 1 kilogram weight, and so 1 newton is equivalent to $\frac{1}{9.81}$ kilogram weight, i.e. 102 gram weight.

PASCALS, BARS

Pressure is the force applied or distributed over a surface, and is expressed as force per unit area. The SI unit of pressure is the pascal (Pa) and 1 pascal is a pressure of 1 newton acting over an area of 1 square metre. As the weight of 102 grams acting over 1 square metre represents a tiny pressure, the unit of pressure commonly used is not the pascal but the kilopascal (kPa). For high pressure gas supplies the bar is used as the unit. Although it is not an SI unit it has been retained for general use. One bar equals 100 kilopascals and is also the approximate atmospheric pressure at sea level. Consequently, the gauge pressure of 137 bar in the full oxygen cylinder is equivalent to about 137 atmospheres. The accurate inter-relationship of bars and atmospheres is considered further in Chapter 4.

THE INTER-RELATIONSHIP OF PRESSURE AND FORCE

It is easier to understand pressure and force in the context of examples taken from anaesthetic practice.

Figure 1.1 illustrates the relative difficulty of injecting a liquid from a large and from a small syringe. The pressure developed in the syringe depends on the force and the area over which it is applied.

$$P = \frac{f}{a}$$

where P = pressure
f = force
a = area

If the force exerted by the thumb is similar for the two syringes the pressure available for injection is greatly increased by the small area of the plunger of the small syringe, because this pressure is inversely proportional to the cross-sectional area of the plunger. If this area is increased by a factor of four by doubling the syringe diameter, then the pressure generated is reduced by a factor of four, provided that the force on the plunger is the same.

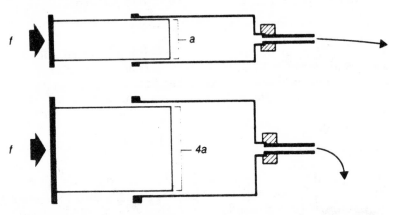

Figure 1.1 **Effect of doubling syringe diameter on the pressure which is generated.**

Thumb pressure on the end of a syringe plunger can produce a force of 25 newtons. If the area of the plunger in a 2 ml syringe is 5×10^{-5} square metres, the pressure generated in the syringe is as follows:

$$\frac{25\,\text{N}}{5 \times 10^{-5}\text{m}^2} = 500\,\text{kPa}$$

The pressure of 500 kPa is approximately five times atmospheric pressure, so it is easy to produce extravascular infusion unintentionally with such a syringe.

The area of the plunger of a 20 ml syringe is greater, e.g. $2·5 \times 10^{-4}$ square metres, so a similar calculation shows that a lower pressure of 100 kPa (about one atmosphere) can be produced. Nevertheless, this is still six times a typical systolic blood pressure of 16 kPa (120 mmHg). The pressure generated when local analgesic is injected from a 20 ml syringe can give rise to accidents during a technique known as intravenous regional analgesia. In this technique a blood pressure cuff inflated to above systolic pressure is used to protect the patient from the systemic effects of local analgesic injected into a distal vein. From the calculations already given it can be seen that pressures in the vein during rapid injection can rise to over the systolic pressure, so the protection from the blood pressure cuff can be inadequate, particularly if a vein adjacent to the cuff is used.

Some syringe pumps and infusion pumps can also generate very high pressures, and care should be taken to ensure that cannulae are correctly placed and that extravascular infusion does not occur.

Another clinical example of pressure generated by force over an area is the formation of bed sores in an immobilised patient. Suppose 20 kg of the patient's weight are supported by an area of contact between the patient and the bed of $10^{-2} m^2$ (an area equal to 10 cm $\times$ 10 cm). The force over this area is:

$$20 \text{ kg} \times 9·81 \text{m s}^{-2} = 196 \text{ N}$$

The pressure is therefore:

$$\frac{196 \text{ N}}{10^{-2} m^2} = 19·6 \text{ kPa}$$

A typical systolic blood pressure is only 16 kPa, so the blood supply to this area is cut off and there is a risk of ischaemia and bed sores at this pressure point.

A pressure relief valve and the expiratory valve of anaesthetic systems also provide simple examples of the balance between force and pressure exerted over an area.

As shown in Fig. 1.2, the pressure P inside an anaesthetic delivery system acts over the area a of the disc of an expiratory valve to exert a force. If this force is greater than the force f exerted by the spring, the disc valve rises to release gas. In the case of the expiratory valve a very light spring is used so that low pressures (e.g. 50 Pa) suffice to open the valve at its minimum setting, but the anaesthetist can raise the pressure by screwing down the cap above the spring, if required.

In the safety valve found on most anaesthetic machines a stronger

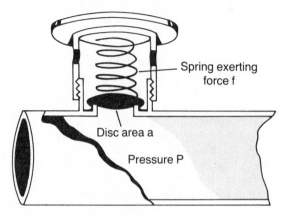

Figure 1.2   Pressure relief valve.

spring is used so that the pressure cannot rise above about 35 kPa and damage the component parts of the machine. A similar safety valve is often incorporated into ventilators but is set to a lower level (e.g. 7 kPa).

Pressure-reducing valves are used in the anaesthetic machine to reduce the pressure and control the supply of gas from the cylinders. These valves also work by balancing the force of a spring against that from the pressure on a diaphragm.

In the pressure-reducing valve illustrated in Fig. 1.3, as the low pressure P_2 falls, the force acting on the diaphragm from below falls and the spring pushes the diaphragm down. When the diaphragm descends

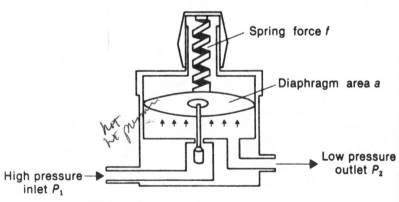

Figure 1.3   Pressure-reducing valve.

it carries with it a rod connected to a small valve which controls the supply of gas at high pressure P_1 and so maintains the pressure P_2 in the compartment below the diaphragm at its correct level.

Figure 1.3 shows a single-stage reducing valve, but two-stage valves are also available.

Figure 1.4 shows an Entonox valve, a two-stage valve in which the first stage is identical to the reducing valve described above. The flow of gas from the outlet of the second-stage valve is controlled by a large diaphragm 'd'. Movements of this diaphragm tilt a rod which regulates the flow of the gas out of the first-stage valve. The second stage is adjusted so that gas flows only when the pressure is below atmospheric. The Entonox valve is an example of a demand valve and similar valves are sometimes used by firemen in breathing apparatus and by aeroplane pilots for oxygen administration.

Control and warning mechanisms which are attached to the high pressure oxygen supply in many anaesthetic machines operate on a similar principle to the reducing valve.

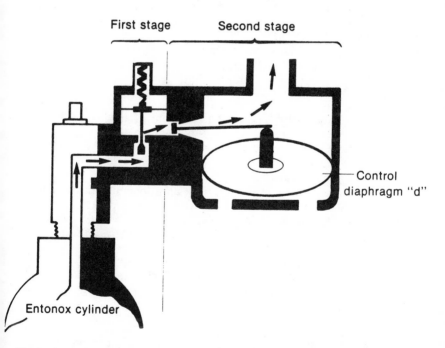

Figure 1.4 Entonox valve, illustrating the first-stage and second-stage pressure-reducing valves.

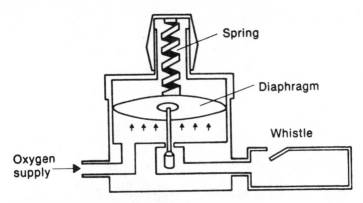

Figure 1.5   Oxygen-failure warning device.

An example of an oxygen-failure warning device is illustrated in Fig. 1.5. When the oxygen supply pressure falls below a set minimum value, the force acting on the diaphragm fails to hold it against the spring. The diaphragm moves down and oxygen leaks past the small valve to blow the whistle, thereby giving a warning of the falling oxygen pressure.

A system of this type with a diaphragm and valve may also be used to cut off the nitrous oxide supply during an oxygen supply failure.

GAUGE AND ABSOLUTE PRESSURES

A full oxygen cylinder has a gauge pressure of 137 bar. When the cylinder is empty the gauge records 0 bar but, unless a vacuum pump has been used, the cylinder still contains oxygen at the ambient atmospheric pressure, and the true or absolute pressure in the empty cylinder is about 1 bar (Fig. 1.6).

In most cases, anaesthetists can ignore atmospheric pressure and use gauges that record the gauge pressure above or below existing atmospheric pressure. Thus, ventilator and gas-cylinder pressures, arterial blood pressure and venous pressure readings are all gauge pressures.

To avoid confusion the term 'absolute pressure' is used when the total pressure including atmospheric pressure is required. The absolute pressure for the empty oxygen cylinder is about 1 bar; so for a full cylinder, if the gauge pressure is 137 bar the absolute pressure is 138 bar.

Absolute pressure = Gauge pressure + Atmospheric pressure

Atmospheric pressure at the surface of the earth is due to the

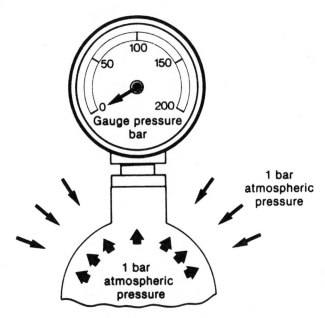

Figure 1.6 Inter-relationship of gauge, atmospheric and absolute pressure.

gravitational force on the air molecules above, the actual pressure depending on the density of air over the point concerned which, in turn, depends on altitude and weather conditions.

PRESSURE IN LIQUIDS

In the same way that the gravitational force on the air molecules gives rise to atmospheric pressure, so too the force of gravity on the molecules of a liquid gives rise to a pressure that depends on the height of liquid above the point of measurement. This is used as the measuring principle in instruments such as manometers and is normally independent of the cross-sectional area of the column of liquid used.

Figure 1.7 illustrates the pressures P_1 and P_2 at the bottom of 10·2 cm high water columns with cross-sectional areas of 1 cm^2 and 4 cm^2, respectively. The pressures P_1 and P_2 can be calculated if it is remembered that 10·2 cm^3 water weighs 10·2 g. Therefore:

$$P_1 = \frac{f_1}{a_1} = \frac{10 \cdot 2 \text{ g wt}}{1 \text{ cm}^2} = 10 \cdot 2 \text{ g wt cm}^{-2}$$

and

$$P_2 = \frac{f_2}{a_2} = \frac{4 \times 10\cdot2 \text{ g wt}}{4 \text{ cm}^2} = 10\cdot2 \text{ g wt cm}^{-2}$$

The example shows that the pressure in water at $10\cdot2$ cm depth will always be $10\cdot2$ g wt cm^{-2}.

To convert the unorthodox figure of $10\cdot2$ g wt cm^{-2} into SI units the calculation is as follows:

$$\text{Pressure of } 10\cdot2 \text{ cmH}_2\text{O} = 10\cdot2 \text{ g wt cm}^{-2}$$
$$= 10\cdot2 \times 10^4 \text{ g wt m}^{-2}$$
$$= 102 \text{ kg wt m}^{-2}$$

since 102 g wt $= 1$ N

$$\text{Pressure of } 10\cdot2 \text{ cmH}_2\text{O} = 1 \text{ kN m}^{-2}$$
$$= 1 \text{ kPa}$$

Hence $10\cdot2$ cm of water pressure is 1 kilopascal. In practice, however, most water manometers are still calibrated in centimetres of water pressure and are calibrated so that the height of the liquid column can be read directly.

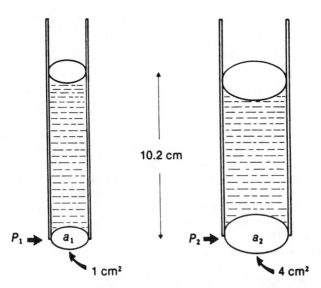

Figure 1.7   Pressure at the base of a column of water.

EFFECT OF SURFACE TENSION

Although the pressure is normally unaffected by the diameter of the manometer tube, surface tension may influence the readings in narrow capillary tubes, increasing the reading in the case of a water manometer and decreasing it in a mercury manometer.

Figure 1.8 illustrates the molecular basis of surface tension. On the left, the diagram shows the forces of attraction which act on a molecule at some depth in a liquid and those which act on a molecule in the surface layer. In the surface layer, some of the forces of attraction between the molecules act in a direction parallel to the surface of the liquid and result in the liquid surface behaving as though a skin were present. This is the phenomenon known as surface tension. If all the forces between the molecules on either side of an imaginary straight line on the surface of the liquid are summed, the resulting surface tension may be expressed in terms of force per unit length (newtons per metre in SI units).

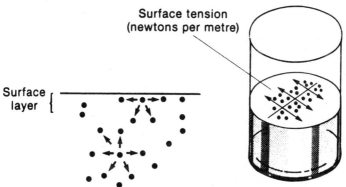

Figure 1.8 Molecular basis of surface tension.

In addition to the forces of attraction between the molecules of a liquid, there are also forces between the molecules and the walls of the containing vessel. These forces result in the meniscus, the curvature of a liquid surface at its contact with the wall. Figure 1.9 shows a vertical glass capillary tube with its lower end immersed in water. Because of the meniscus, the surface tension (T) at the junction between the water and the wall of the tube acts to some extent in a vertical direction and pulls the water up the tube until it is balanced by the force of gravity on the water in the tube which is above the normal liquid level. The water rises up the capillary tube to a height that is inversely proportional to the radius of the tube.

In the case of mercury, the meniscus is convex and so surface tension depresses the level of liquid in the capillary tube.

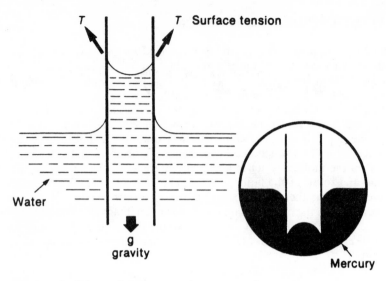

Figure 1.9   Effect of surface tension at a meniscus.

EFFECT OF SLOPE IN A MANOMETER TUBE

If a manometer tube is tilted, the liquid rises along it until the necessary vertical height is achieved (Fig. 1.10). Particularly sensitive gauges sometimes use a sloped manometer tube and are appropriately calibrated, but in anaesthetic practice most manometer tubes are intended for vertical use.

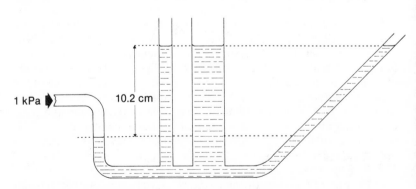

Figure 1.10   Effect of slope on manometer calibration.

MERCURY MANOMETERS

If a dense liquid such as mercury is used, a manometer may be used to measure higher pressures. Mercury is 13·6 times as dense as water, so the force exerted by its weight is proportionately greater. A pressure which supports a 7·5 mm column of mercury will support a 10·2 cm column of water (Fig. 1.11).

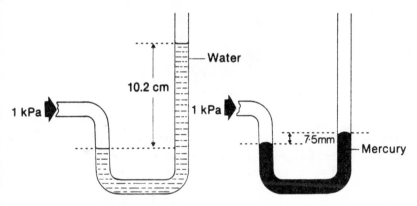

Figure 1.11   Comparison of water and mercury manometers.

The pressures in the two manometers may be compared as follows:

$$\text{Pressure exerted by } 7\text{·}5 \text{ mmHg} = 13\text{·}6 \times 7\text{·}5 \text{ mmH}_2\text{O}$$
$$= 10\text{·}2 \text{ cmH}_2\text{O}$$
$$= 1 \text{ kPa}$$

Mercury is useful in manometers measuring up to atmospheric pressure (1 bar or 100 kPa) and, as 1kPa equals 7·5 mmHg, it is obvious that 1 bar is 750 mmHg pressure. The units of bar and kilopascal have the advantage that they are independent of changes of gravitational force and of temperature, which can affect the height of a mercury column.

It is essential that the top of a manometer tube is open. In the mercury manometers used for blood pressure measurements, a disc of material permeable to air is placed at the top of the tube to prevent spillage of the mercury, but this creates a potential problem in that dirt or grease may partially obstruct the disc thus leading to faulty readings.

In contrast to the standard open-ended manometer tube used for gauge pressure, the mercury barometer used to measure atmospheric pressure is sealed, as shown in Fig. 1.12. A vacuum is present above the mercury so that the full absolute pressure is recorded.

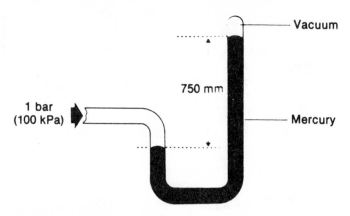

Figure 1.12 Mercury barometer.

In some conditions in anaesthetic practice, pressures much higher than 1 bar are present and in these cases a Bourdon gauge (illustrated in Fig. 1.13) is used. In this gauge, the gas at high pressure causes a tube to uncoil and in doing so moves a pointer over a scale on a dial. Bourdon gauges have the advantage over manometers that there is no liquid to spill, and they are sometimes called aneroid gauges from the Greek 'a-neros' (without liquid).

Another form of aneroid gauge is based on a bellows or capsule which expands or contracts depending on the pressure across it. Two

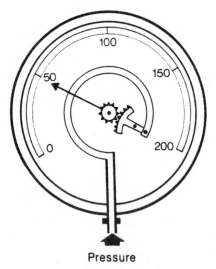

Figure 1.13 Bourdon gauge.

spring loaded.

same as aneroid

forms are illustrated in Fig. 1.14: the one on the left is sealed and is
sensitive to changes of atmospheric or absolute pressure; the one on the
right has the interior open to the surrounding air and is suitable for
measuring gauge pressure.

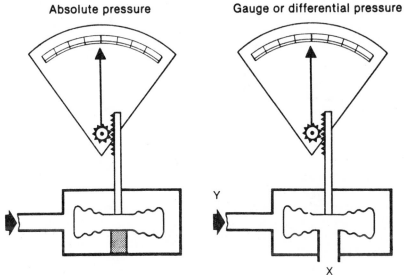

Absolute pressure   Gauge or differential pressure

Figure 1.14   Types of bellows aneroid gauges.

DIFFERENTIAL PRESSURE

In addition to absolute and gauge pressure, differential pressure is sometimes measured. Differential pressure is the difference between two
pressures, e.g. between two points in an anaesthetic breathing system.
The aneroid gauge shown on the right of Fig. 1.14 can act as a
differential gauge if connections are made at X and Y, i.e. to the inside
and outside of the capsule.

OTHER UNITS

In this chapter the units pascal, bar, mmHg and cmH_2O have been used
because these are the units encountered in anaesthetic practice. On
older anaesthetic machines the pressure gauges may be calibrated in
$kg\ wt\ cm^{-2}$ but conversion is easy as $1\ kg\ wt\ cm^{-2}$ is approximately
1 bar (100 kPa). Pounds per square inch too may still be seen on some
machines and $14.5\ lb\ inch^{-2}$ is 1 bar. Torr may be seen on vacuum
pressure gauges and 7·5 torr is 1 kPa (1 torr is about 1 mmHg).

Fluid Flow

Flow is defined as the quantity of a fluid, i.e. a gas or a liquid, passing a point in unit time.

$$F = \frac{Q}{t}$$

where F = mean flow
$\quad Q$ = quantity (mass or volume)
$\quad t$ = time

In this chapter, flow is represented by the rate of change of quantity to avoid the need to specify mass or volume. On graphs and illustrations the symbol $\dot{Q}$ is used, the small dot above the Q indicating rate of change of Q. The symbol Q without the dot stands for quantity.

LAMINAR FLOW

In laminar flow (Fig. 2.1) a fluid moves in a steady manner and there are no eddies or turbulence. This is the type of flow normally present in smooth tubes at low rates of flow.

The flow is greatest in the centre, being about twice the mean flow, as illustrated by the longer arrows in the figure. As the side of the tube is

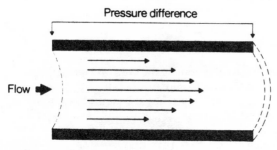

Figure 2.1   Laminar flow.

approached the flow becomes slower until it approaches zero at the wall. In order to drive a fluid through a tube, a pressure difference must be present across the ends.

The graph in Fig. 2.2 shows the result if various flows are passed through a tube and the resulting pressure drop across the ends is recorded. There is a linear relationship so that flow is directly proportional to pressure under conditions of laminar flow. The ratio of pressure to flow is a constant known as the resistance R of the apparatus or the tube concerned.

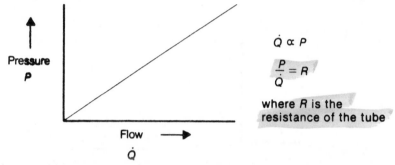

Figure 2.2 The graph shows the linear relationship of flow to pressure in laminar flow.

Figure 2.3 shows how resistance can be measured. A constant flow $\dot{Q}$ is passed through the apparatus concerned and the difference in pressure $P_1 - P_2$ between the ends of the apparatus is measured. By dividing the pressure difference by the flow the resistance of the apparatus is obtained and, provided that flow is laminar, the resistance is independent of the flow. A technique such as this may be used to measure resistance with either gas or liquid flow. For liquid flow a constant head of pressure may be achieved by means of a reservoir of

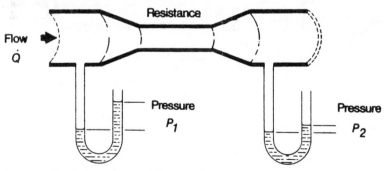

Figure 2.3 Measurement of flow-resistance.

liquid and measurement of the resulting flow allows calculation of the resistance. A system of this type may be used to illustrate what happens if the radius of a tube is halved (Fig. 2.4).

It is found that this alteration of radius has a pronounced effect on resistance to flow. Halving the radius reduces the flow to one-sixteenth of its original value if the pressure drop along the tube remains the same. In other words, the flow is proportional to the fourth power of the radius. Consequently, a slight reduction of the radius of an endotracheal tube can have an appreciable effect on resistance and therefore on flow.

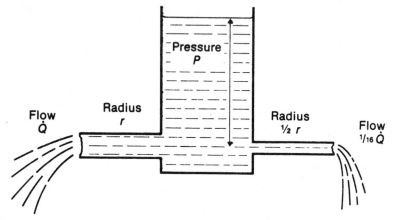

Figure 2.4   Effect on flow of halving the tube radius.

As shown in Fig. 2.5, the effect of altering the length of the tube is much less marked than that of altering the radius. If the length of the tube is halved, the flow will double, other factors being kept constant. Finally, the viscosity of the fluid affects resistance to laminar flow in such a way that the higher the viscosity the slower is the flow. Viscosity is a measure of the frictional forces acting between the layers of the fluid as it flows along the tube. It is represented by the Greek letter eta (η) and has the units of pascal seconds.

The effects described may be summarised as follows:

$$\dot{Q} \propto P$$

$$\propto r^4$$

$$\propto \frac{1}{l}$$

$$\propto \frac{1}{\eta}$$

where $\dot{Q}$ = flow through tube
  P = pressure across tube
  r = radius of tube
  l = length of tube
  η = viscosity of fluid

All these factors are incorporated in the equation known as the Hagen-Poiseuille equation as follows:

$$\dot{Q} = \frac{\pi P r^4}{8\eta l}$$

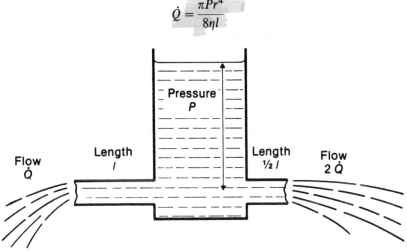

Figure 2.5 Effect on flow of halving the length of a tube.

TURBULENT FLOW

Laminar flow is not the only type of flow occurring in anaesthetic apparatus and in the circulation.

As illustrated in Fig. 2.6, laminar flow may change to turbulent flow if a constriction is reached which results in the fluid velocity increasing. In turbulent flow, fluid no longer flows in a smooth fashion but swirls in eddies. The difference in fluid velocity across the tube that is present in laminar flow no longer occurs in the case of turbulent flow. Also the flow is no longer directly proportional to the pressure as in the case of laminar flow.

The relationship is shown in Fig. 2.7, the flow being approximately proportional to the square root of the pressure. Thus, in order to double the flow in a given piece of apparatus, the pressure needs to be raised by a factor of four. As the relationship of pressure to flow is no longer linear, resistance is not constant and when referring to resistance in the

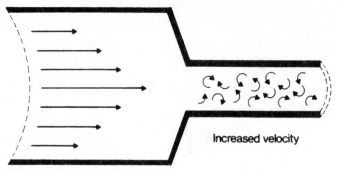

Figure 2.6 Turbulent flow.

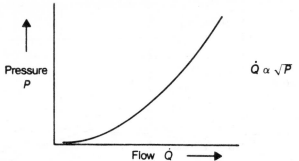

Figure 2.7 The graph shows the relationship of flow to pressure in turbulent flow.

presence of turbulent flow it is important that the flow, at which the resistance is measured, is specified—for example, the air flow resistance in breathing depends on the air flow.

The other factors affecting turbulent flow may be summarised as follows:

$$\dot{Q} \propto \sqrt{P} \text{ approx}$$

$$\propto r^2$$

$$\propto \frac{1}{l}$$

$$\propto \frac{1}{\rho}$$

where $\dot{Q}$ = flow
r = radius of tube
l = length of tube
ρ = density of fluid

Turbulent flow varies approximately with the square of the radius. The fact that the radius has less effect on turbulent flow than laminar flow is to be expected, because in the case of turbulent flow the flow is more evenly distributed over the diameter of the tube. Like laminar flow, turbulent flow is affected by the length of the tube and flow increases with a shorter tube. The characteristics of the fluid passing through the tube also have an effect but it is the density which is important in turbulent flow, density being represented by the Greek letter rho (ρ) and being equal to mass divided by volume (kg m^{-3}).

ONSET OF TURBULENT FLOW

As illustrated in Fig. 2.6, turbulent flow may occur if there is a sharp increase in the flow through a tube, but there are several other factors influencing the type of flow, such as the viscosity and density of the fluid and the diameter of the tube. These factors may all be combined to give an index known as Reynold's number. Reynold's number is calculated from the factors as follows:

$$\text{Reynold's number} = \frac{v\rho d}{\eta}$$

where v = linear velocity of fluid
 ρ = density
 d = diameter of tube
 η = viscosity

Empirical measurements with cylindrical tubes show that, if Reynold's number exceeds about 2000, then turbulent flow is likely to be present. If Reynold's number is below 2000, then the flow is usually laminar.

It can be seen from this formula that for a fixed set of conditions there is a critical velocity at which Reynold's number has the value of 2000. When the velocity of the fluid exceeds this critical value, the character of the flow is likely to change from laminar to turbulent. This critical velocity applies only for a given fluid in a given tube. An example of the effect of density on the onset of turbulent flow is the use of helium in respiratory disorders. In this case helium reduces the density of gas inhaled and so reduces the incidence of turbulent flow. A lower resistance to breathing results because the transition from turbulent to laminar flow is associated with a marked decrease in resistance to flow through a tube.

Turbulent flow is often present where there is an orifice, a sharp bend or some other irregularity.

CLINICAL ASPECTS OF FLOW

Changes between laminar and turbulent flow depend on the velocity of the gas, which in turn depends on the volume flow and thus on the diameter of the tubing and airways. The flow at which transition takes place depends on the gases present, and Fig. 2.8 illustrates the critical flows for a typical anaesthetic mixture of 60% nitrous oxide in oxygen and also for air at an ambient temperature of 20 °C. In the patient's airway, gases are humidified, contain carbon dioxide and are at a temperature of 34 to 37 °C. The overall effect of these changes is a rise in critical flow due to the dominant effect of the reduction in density from the warming of the gases. Consequently, the critical flow for the humidified warmed anaesthetic mixture in the patient could be 10% higher than the figures for dry gases at 20 °C illustrated on the graph.

As shown in Fig. 2.8, the critical flow in litres per minute for a typical anaesthetic gas has approximately the same numerical value as the diameter of the airway concerned in millimetres. Thus the flow of anaesthetic gases in a 9 mm internal diameter endotracheal tube will become turbulent when flow exceeds about 9 litres per minute (9 litre min^{-1}), the flow in a 15 mm diameter trachea is turbulent at flows over 15 litres per minute (15 litre min^{-1}), while in the 22 mm internal diameter tubing of breathing systems flow becomes turbulent when flows over 22 litres per minute (22 litre min^{-1}) occur. As breathing is

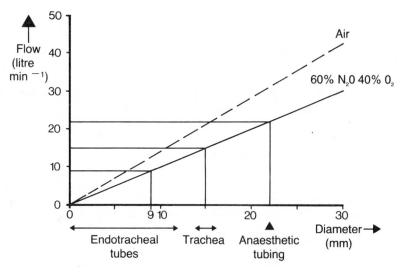

Figure 2.8   Critical flow in smooth tubes for air and for a 60% nitrous oxide 40% oxygen mixture. The values are for dry gas at 20 °C and would be slightly higher for warmed humidified gas (see text).

cyclical, with peak flows over 50 litre min^{-1}, turbulent flow usually predominates during peak flow while laminar flow is present at other times in the respiratory cycle. As shown in Fig. 2.8, the critical flows for air are higher than those for a nitrous oxide–oxygen mixture, so laminar flow prevails to a greater extent, especially in the wider breathing tubes.

Although the bronchi and smaller air passages in the lungs are even narrower than the trachea the air flow through them is slower, so laminar flow is usual in the lower respiratory tract.

The examples given assume that the tubes are smooth. The corrugated surface of anaesthetic tubing induces turbulence at a lower flow than a smooth tube. In addition, any sharp bends or angles increase the incidence of turbulent flow. As an example, Fig. 2.9 illustrates turbulence at the connector of an endotracheal tube.

In quiet breathing, the flow within the respiratory tract is principally laminar, but during speaking, coughing, or taking a deep breath, turbulent flow tends to occur as the critical velocity is exceeded. In practice, many other factors can increase the likelihood of turbulent flow in the respiratory tract, e.g. a lining layer of mucus may affect the flow by narrowing the respiratory passages.

Similar considerations apply to the circulation, the flow being principally laminar, but with the possibility of turbulence at the junctions of vessels or where vessels are constricted by outside pressures. Turbulence in the circulatory system often results in a bruit which can be heard on auscultation.

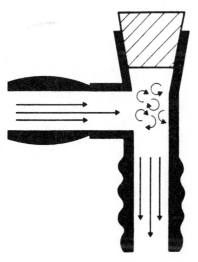

Figure 2.9 Turbulent flow at the connector for an endotracheal tube.

The blood vessels and bronchi, however, are not rigid tubes but have a variable radius depending on the pressure gradient across their walls and on the tone of the elastic and smooth-muscle tissue in their walls. This tone gives rise to tension in the tubular wall and changes in tension by altering the radius thus affect flow.

TENSION

As shown in Fig. 2.10, tension is a tangential force in newtons per metre acting on a length of the wall A to B. A balance must be present between the pressure caused by the smooth muscle and elastic tissue and the fluid pressure in the tube in order to prevent the tube progressively distending or collapsing. The pressure caused by the tension of the smooth muscle and elastic tissue may be calculated from Laplace's law. For a tube, Laplace's law states:

$$\text{Pressure gradient across the wall of a tube} = \frac{\text{Tension}}{\text{Radius}}$$

The application of this law shows a potential instability in the radius of vessels. A fall in pressure in an arteriole, for example, tends to distend it less and so could reduce its radius, but the smooth muscle in the wall maintains tension and so the ratio tension to radius is increased and pressure across the wall is raised (Laplace's law). Instability then results with contraction and closure of the vessel. This can occur clinically at a

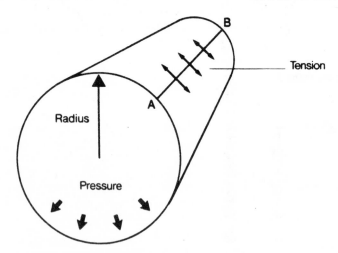

Figure 2.10   Inter-relationship between pressure, radius and tension in a distensible tube.

low pressure known as the critical closing pressure for the vessel. Similarly, bronchioles and alveolar ducts also may close when a critical closing pressure is present in the airways on expiration. For vessels and bronchi to remain patent a correct balance between pressures across their walls and the tensions from the elastic tissue and smooth muscle is essential. In addition to these sources of tension, surface tension is also present at air–liquid interfaces as explained in Chapter 1.

SURFACE TENSION

The surface tension in alveoli and alveolar ducts can lead to potential instability. When the radius of the alveolar ducts and alveoli diminish during expiration, the pressure across the wall tends to rise (Laplace's law) and would lead to collapse of the alveoli if the surface tension were to remain constant. In the case of the spherical alveoli the pressure generated by the surface tension is double that in the tubular ducts as there are two planes of curvature. Laplace's law for a spherical shape is thus:

$$\text{Pressure gradient across the wall of a sphere} = \frac{2 \times \text{Tension}}{\text{Radius}}$$

Surfactant, present in the lining fluid of the alveoli, protects the patient from this danger by two mechanisms. It reduces appreciably the surface tension of the fluid lining the alveoli, and also renders the surface tension variable so that the tension increases as the alveoli distend and decreases as the alveoli contract, thus preventing collapse. In premature infants, surfactant may not yet be present in the alveoli and consequently problems may arise because of alveolar collapse and failure of alveoli to open.

VISCOSITY

In addition to changes of vessel and airway diameters due to tension in the walls, viscosity of the contained fluid, especially if liquid, may also affect flow. In the circulation, for example, blood viscosity can vary and is increased if a patient's haemoglobin or fibrinogen level is raised. A rise in viscosity reduces blood flow, giving a risk of vascular occlusion. Viscosity increases at low temperature, with patient age and with cigarette smoking while treatment with low molecular weight dextran reduces it. The viscosity of blood is anomalous because blood is not a uniform fluid but contains cells. This leads to a rise in the viscosity of blood at very low flows such as those which occur in areas with intermittent circulation.

A viscometer is used to measure viscosity. In a typical viscometer a sample of blood is placed in a cone-shaped cup which is rotated at a standardised rate and maintained at a controlled temperature (Fig. 2.11). A cone-shaped probe within the cup is supported on a torsion wire and is dragged by the viscosity of the blood to turn a mirror fixed to its axis. This alters the position of the light beam on a scale, thus providing a measure of the viscosity. Other viscometers determine viscosity by measuring the time taken for a sample of liquid to flow down a vertical tube. The transit time is measured by detecting the meniscus position with photocells and the viscosity can then be computed electronically from this transit time.

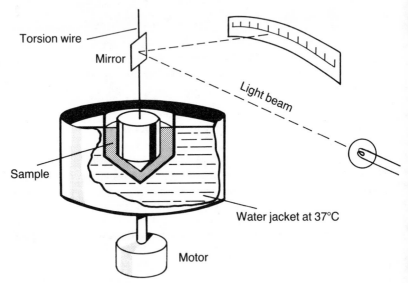

Figure 2.11   Viscometer.

FLOW THROUGH A VENTURI

Some equipment may include a steadily widening tube called a venturi, illustrated in Fig. 2.12. The flow in such a tube remains laminar if the increase in cross-section is gradual. If the pressure along the tube is measured, it is found that the pressure at the narrowest point is lower than elsewhere and is often below atmospheric pressure. This fall of pressure at the point of narrowing of a tube is named after its discoverer, Bernoulli, and is applied in several devices.

The fall in pressure at the narrowing of a tube arises in the following

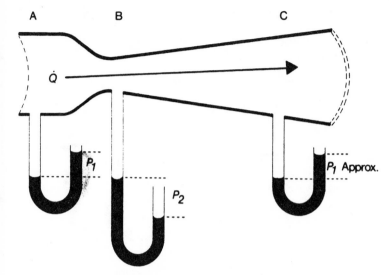

Figure 2.12 Principle of the venturi.

way. Flowing fluid contains energy in two forms, potential energy associated with its pressure and kinetic energy associated with its flow. At a point of narrowing, B in the diagram, there is a considerable increase of fluid velocity and a great gain of kinetic energy. Such an increase of kinetic energy can only occur if there is a fall in potential energy, because the total energy present must remain constant. In consequence, a marked fall of pressure occurs at this point to pressures below those at A and C. As the tube widens the fluid velocity falls and with it the kinetic energy component, so that the potential energy, and hence the pressure, rise. For the system to work efficiently, laminar flow is important as turbulence would allow flow energy to be lost as heat.

Consider now what happens if, instead of a pressure gauge at position P_2, the side tube is open. If pressure at this point is below atmospheric, air or other fluid can be entrained through the side arm. Such an apparatus is called an injector, and Fig. 2.13 shows how an injector with its venturi operates.

The driving fluid entrains fluid through a side arm because of the low pressure at the point where the fluid velocity is greatest—just at the orifice. Injectors working on this principle may be used to provide suction, either water or oxygen being used as the driving fluid.

A nebuliser (Fig. 2.14) is another example of this principle. In this case, gas as the driving fluid enters by the centre tube, entrains liquid from a side tube and breaks it up into droplets suitable for inhalation. Nebulisers are discussed more fully in Chapter 12.

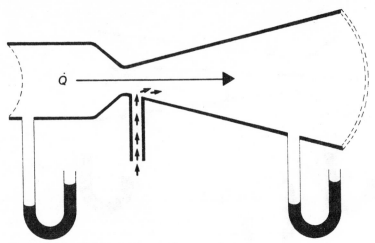

Figure 2.13   Fluid entrainment by a venturi.

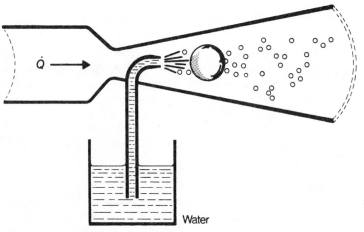

Water

Figure 2.14   Principle of a nebuliser.

Injectors are sometimes used in ventilators to dilute the driving gas, often oxygen, with air or other gases. Such a system is shown in Fig. 2.15. A cycling device is interposed between the venturi and the patient's lungs to complete the simple ventilator.

The oxygen mask illustrated in Fig. 2.16 also incorporates an injector. This injector acts as a system to dilute oxygen with air thus providing clinically useful concentrations. This principle is used in other types of oxygen therapy apparatus and in other inhalational apparatus.

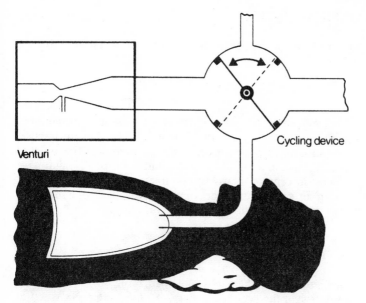

Figure 2.15 Use of a venturi as a source of driving gas in a ventilator.

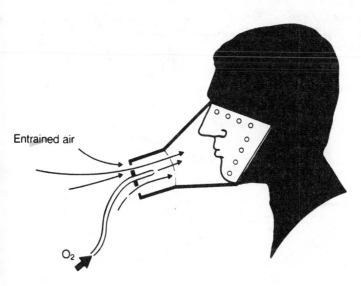

Figure 2.16 Venturi oxygen mask.

ENTRAINMENT RATIO

The entrainment ratio is defined as the ratio of entrained flow to driving flow.

$$\text{Entrainment ratio} = \frac{\text{Entrained flow}}{\text{Driving flow}}$$

Thus, a 9 to 1 entrainment ratio indicates that there are 9 litre min^{-1} being entrained by a driving gas of 1 litre min^{-1}. In clinical practice, entrainment ratios may not be constant and obstruction at the outlet of the venturi can result in a fall in the entrainment ratio. This may give rise to problems in ventilators worked by injectors in which oxygen is used to entrain air or nitrous oxide. In such cases, back pressure may alter the flow and so the resulting oxygen concentration.

THE COANDA EFFECT

As explained previously, there is a fall in pressure in fluid flowing through a tube when the tube narrows due to the increase in speed of the stream of fluid. This low pressure can be used to entrain other fluid into the stream. If such a stream flows along a solid surface such as the wall of a wide tube, air cannot be entrained where the stream touches the wall, so a region of low pressure remains in this area and the stream is held against the wall. Figure 2.17 illustrates this effect in a system where a narrow tube enters a Y-junction of wider bore and it is seen that, because the flow tends to cling to one side, the flow does not divide evenly between the two outlets but flows through only one limb of the 'Y'. This behaviour is known as the Coanda effect and may be of importance in explaining the maldistribution of gas flow to alveoli where there has been a slight narrowing of the bronchiole before it

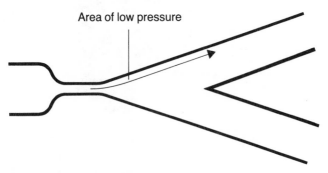

Area of low pressure

Figure 2.17   Coanda effect.

divides. It may also explain some cases of myocardial infarction, where there may be narrowing before the branching of a coronary artery. In such a case all the blood could be flowing into one limb of the coronary artery despite the fact that there is still an apparently patent lumen in the other branch.

A valve mechanism with no moving parts can be made if two tubes are inserted at each side at the exit of the narrow tube (Fig. 2.18). Flow through either of these side tubes can switch the main flow from one exit tube to the other. Having been transferred, the flow continues into the appropriate branch after the switching flow has been removed. Devices of this kind are known as fluid logic, and may be incorporated in ventilators to reduce the number of valves and moving parts. However, such ventilators have the problems of being rather wasteful of gas, and are also often noisy.

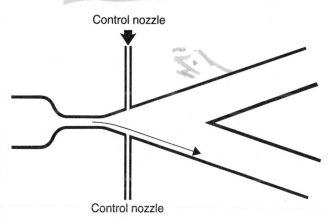

Control nozzle

Control nozzle

Figure 2.18   Valve mechanism based on the Coanda effect.

CHAPTER
3

Volume and Flow Measurement

Volume and flow are related, for, if the volume of a fluid passing some point can be measured over a given time, the flow can be calculated. As gases are compressible there are special problems associated with the measurement of gas flows and volumes which are not present in the case of liquids.

GAS VOLUME MEASUREMENT

The Benedict Roth spirometer (Fig. 3.1) is widely used for both physiological and clinical studies. A light bell moves with the patient's

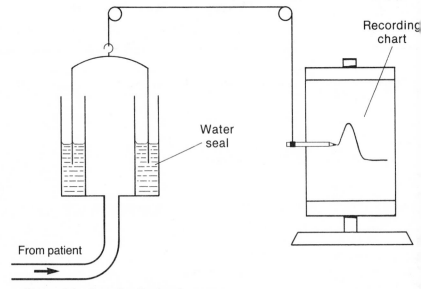

Figure 3.1   Benedict Roth spirometer.

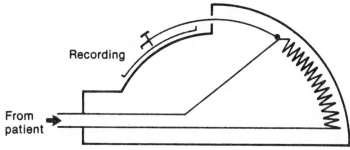

Figure 3.2   Principle of the Vitalograph.

breathing and this movement may be recorded by a pen on a rotating drum, the motion of the bell being transferred to the pen through a connecting wire which passes over two pulleys. A water seal prevents the leakage of gas from the bell and this seal is small in order to reduce the volume of gas which dissolves in the water.

An alternative system for respiratory measurements is the Vitalograph, which is more portable than the Benedict Roth spirometer.

In the Vitalograph (Fig. 3.2), bellows are used to measure gas volume. The top plate of the bellows is pivoted and its motion is transferred to a scriber which records volume changes on a chart. The chart is motor driven and automatically starts to move as soon as the bellows move from their empty position. This allows expired volume–time graphs to be plotted.

Both these spirometers are suitable for measuring limited gas volumes up to a few litres. Occasionally in anaesthesia, larger volumes must be measured and in this case a dry gas meter may be used. The instrument is similar to that used in domestic gas supplies and a simplified meter is illustrated in Fig. 3.3. In this diagram two

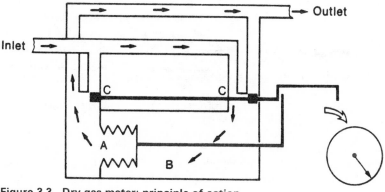

Figure 3.3   Dry gas meter: principle of action.

compartments are shown, A and B. In the position illustrated, gas passes into compartment B compressing the bellows which simultaneously empty compartment A. As the bellows are compressed they move a lever to provide a volume recording and also move a rod CC to shift the valves to a new position (Fig. 3.4).

In this position, compartment A is now filled until it too moves the rod and lever mechanism. The system described is a simplified one and, in fact, two sets of these compartments are necessary for the meter to work continuously. There is also a set of gears to record the volume measured.

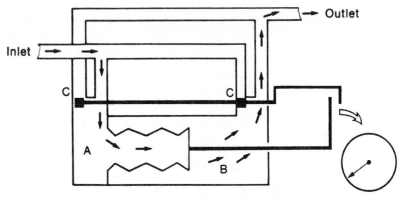

Figure 3.4   Dry gas meter after movement of rod CC.

The compact Wright respirometer is more convenient for measuring tidal volumes in anaesthesia. Figure 3.5 illustrates the principle of its action. Like the dry gas meter it has a set of gears but in this case the volume measurement is achieved by monitoring the continuous rotation of a vane as it is moved by the flow of gas. The effect of the slits is to cause a circular motion of the airflow which rotates the vane. The vane does not rotate when the flow is reversed. The Wright respirometer is calibrated for use for tidal volume measurement and for tidal ventilation. Its calibration is inaccurate if it is used to measure a continuous flow.

An electronic Wright respirometer has been developed which removes the need for the mechanical gears (Fig. 3.6). Mounted above the vane is a special disc with alternating clear and opaque segments and this disc lies between a lamp and a photocell. As the vane rotates, so does the disc, and the photocell is exposed to the light via the clear segments twice per rotation. The photocell converts the pulses of light into electrical pulses, the number of pulses being proportional to the

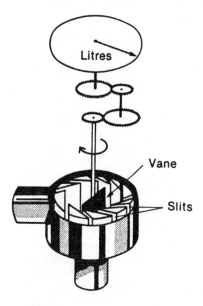

Figure 3.5   Wright respirometer.

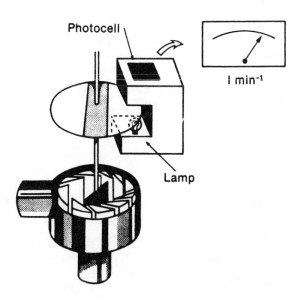

Figure 3.6   Electronic Wright respirometer.

volume of gas flowing. The pulses then pass via a connecting lead to a display unit where they are electronically processed to give a reading, either as tidal volume or minute volume.

This instrument gives more accurate measurements than the non-electronic Wright respirometer because the drag of the gears on the vane is eliminated, but it requires a source of electrical power and is not as portable or cheap as the simpler instrument.

In clinical practice the gas volume measurements made with the instruments described above are often at ambient temperature and pressure. For scientific and research work, however, volumes are corrected to a standard temperature and pressure (s.t.p.) as described in Chapter 4.

LIQUID VOLUME MEASUREMENT

Liquid volume measurement can often be made simply by collecting the volume in a measuring cylinder. For measurement of blood volume in the body more advanced techniques are required, such as those which use the dilution of an indicator, but it is necessary to estimate both the red cell and the plasma volumes.

DILUTION TECHNIQUES

For red cell volume measurement, red cells which have been specially identified by radioactive labelling are used (Chapter 24). Suppose a known amount or dose of these labelled red cells is injected into a patient. After they have mixed thoroughly with the patient's blood the concentration of the labelled cells may be deduced from the radioactivity of a blood sample from the patient.

$$\text{Concentration of labelled red cells} = \frac{\text{Dose of labelled red cells}}{\text{Red cell volume}}$$

Therefore:

$$\text{Red cell volume} = \frac{\text{Dose of labelled red cells}}{\text{Concentration of labelled red cells}}$$

For measuring plasma volume the same dilution technique is used with radioactive albumin instead of radioactive labelled red cells. The total blood volume can then be calculated by adding together the plasma and red cell volumes. It is possible to estimate blood volume from either of these determinations provided that the ratio of red cell volume to blood volume is known from a haematocrit reading.

$$\text{Blood volume} = \frac{\text{Red cell volume}}{\text{Haematocrit}}$$

or

$$\text{Blood volume} = \frac{\text{Plasma volume}}{1 - \text{Haematocrit}}$$

Clinically these dilution techniques have limitations, because in shocked patients pooling and stasis of blood may occur in ill-perfused areas and even mixing of the radioactive indicator may not be obtained. A similar technique employing radioactive sodium has been used to measure total extracellular volume.

This dilution concept is also used to measure blood loss. Consider the blood loss monitor illustrated in Fig. 3.7. A washing reservoir contains in this case 35 litre water. Bloodstained swabs from a patient are added to this reservoir and even mixing of the haemoglobin (Hb) is promoted by bubbles of compressed air. A detector consisting of a lamp and photocell measures the concentration of diluted haemoglobin in the washing reservoir. If the patient's haemoglobin concentration is 140 g litre^{-1} and the diluted concentration of haemoglobin in the reservoir is 1 g litre^{-1} what is the patient's blood loss?

$$\text{Concentration of Hb in reservoir} = \frac{\text{Total amount of Hb}}{\text{Volume of reservoir}}$$

Therefore:

$$\text{Total amount Hb in reservoir} = \text{Concentration} \times \text{Volume}$$
$$= 1 \text{ g litre}^{-1} \times 35 \text{ litre}$$
$$= 35 \text{ g}$$

But, in the patient's blood on the swabs:

$$\text{Concentration of Hb in blood lost} = \frac{\text{Amount of Hb in blood lost}}{\text{Volume of blood lost}}$$

Therefore:

$$\text{Volume of blood lost} = \frac{\text{Amount of Hb}}{\text{Concentration of Hb}}$$
$$= \frac{35 \text{ g}}{140 \text{ g litre}^{-1}}$$
$$= 0 \cdot 25 \text{ litre}$$

This example giving the principle of the technique uses two calculations, but in practice the calculations may be carried out electronically in the apparatus to give a direct reading of blood volume loss.

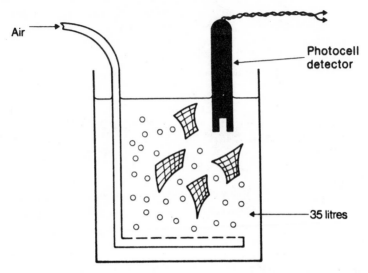

Air

Photocell
detector

35 litres

Figure 3.7   Blood loss monitor. The use of a dilution technique with a washing reservoir to assess blood loss.

GAS FLOW MEASUREMENT

Anaesthetists require a continuous indication of gas flows in the anaesthetic machine and this is provided by a variable orifice flow-meter, often referred to by the trade name of Rotameter. Figure 3.8 illustrates the flowmeter. A bobbin is supported in the middle of a tapered glass or plastic tube by the gas flow and, as the flow increases, the bobbin rises in the tube and the clearance round the bobbin increases. In other words, there is a variable orifice round the bobbin which depends on the gas flow. The pressure across the bobbin remains constant because it gives rise to a force which balances the force of gravity on the bobbin. The increase in the area of the annular orifice as the bobbin rises reduces flow resistance at higher flows and so the pressure across the bobbin stays constant, despite the flow increase.

Small slots are placed round the top of the bobbin as shown, causing it to rotate centrally in the gas flow. In this type of flowmeter, readings are made from the upper surface of the bobbin.

In a variable orifice flowmeter there is a mixture of turbulent and laminar flow, and so for calibration purposes both the density and viscosity of the fluid are important (Chapter 2). Consequently, careful recalibration is required if a flowmeter is used for a different fluid than that for which it was initially designed.

The flowmeter tube must be kept vertical to obtain a correct reading and to prevent the bobbin touching the sides of the tube and sticking. Sticking is more likely to occur when the bobbin is rotating near the bottom of the tube. Electrostatic charges may also build up on the bobbin if it rubs against the side of the tube, and these may increase the tendency to stick. To prevent the build-up of such charges, some tubes have a conductive strip running down the inside at the back, while in others a clear conductive coating is provided inside to conduct away any electrostatic charges.

The problem of sticking is less if a simple ball flowmeter is used as shown in Fig. 3.9, but this may be less accurate because there is no well-

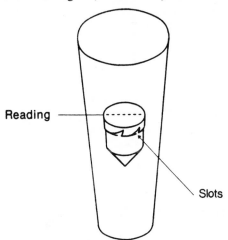

Figure 3.8   Variable orifice flowmeter.

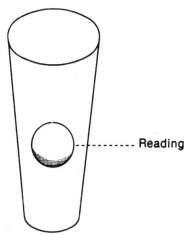

Figure 3.9   Ball flowmeter.

defined surface to read. In practice, the readings are taken from the middle of the ball.

Below the flowmeter there is usually a needle valve, as illustrated in Fig. 3.10. In the needle valve there is a spindle, attached to a control knob which screws into the seating of the inlet to turn off the gas supply to the flowmeter above. Leakage of gas around the spindle is prevented by a gland with its gland nut. A gland is a washer of compressible material, and glands and gland nuts are also used to prevent leakage on gas cylinders. At the bottom of each flowmeter a dust filter of sintered metal is also present.

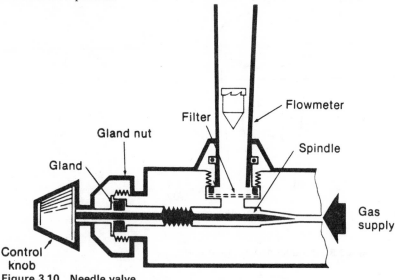

Figure 3.10   Needle valve.

The positions of individual gas flowmeters on the anaesthetic machine may cause problems, as illustrated in Fig. 3.11A. If there is a leak in the centre flowmeter or in the centre of the block, more oxygen (from the left) leaks out through this hole than nitrous oxide, resulting in higher percentages of nitrous oxide than intended being delivered to the patient, perhaps with fatal consequences.

One solution would be to reverse the oxygen and nitrous oxide flowmeter positions as shown in Fig. 3.11B, and these positions may be found in North America. Unfortunately, many anaesthetists are so used to the current position that this solution is not universally acceptable. Therefore, on some machines another solution, illustrated in Fig. 3.11C, is found. Here there is suitable channelling present in order to pick up the oxygen selectively.

A second instrument which measures flow by the variable orifice principle is the Wright peak flowmeter illustrated in Fig. 3.12. In this

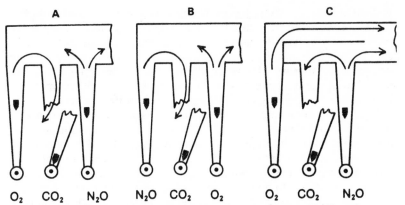

A   B   C

O₂   CO₂   N₂O N₂O CO₂ O₂ O₂   CO₂ N₂O

Figure 3.11. The risk of hypoxia with a faulty flowmeter block and possible design measures to prevent it.

flowmeter the patient's expired gases are directed against a moveable vane causing it to rotate. As the vane rotates, it opens up a circular slot around the base of the instrument and so allows the gases to escape. Rotation of the vane is opposed by the force from a coiled spring and a pointer mounted on the axis of the vane registers its movement on a calibrated dial. The spring force is relatively constant and its action on the area of the vane gives a very small steady pressure to balance the constant pressure driving the flow of gas through the variable orifice. The size of this orifice increases to that required for the gas flow and, at peak flow, the vane reaches a maximum position from which it is

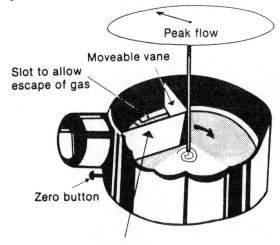

Figure 3.12 Wright peak flowmeter.

prevented from returning by a ratchet. After the maximum reading is taken, the vane can be released by depressing a button and the pointer returns to zero. An alternative, more compact peak flowmeter, has a cylindrical shape and the air escapes from a straight slot in place of the circular one described. However, the variable orifice principle is still used. For an adult a peak flow of 400 to 500 litre min^{-1} is common but in a patient with emphysema this may be below 100 litre min^{-1}.

In research applications in anaesthesia a continuous recording of a patient's respiratory volume and air flow may be needed, and in such cases a pneumotachograph may be used (Fig. 3.13). The measuring head of this instrument contains a gauze screen and has a sufficiently large diameter to ensure laminar flow through the gauze. The gauze acts as a resistance to flow, and so respiratory airflow from the patient causes a small pressure drop across the gauze. This pressure change is measured by a transducer which converts the pressure change into an electrical signal which in turn can be displayed and recorded. (The word transducer describes a device which changes a signal from one form of energy to another.) The pneumotachograph can measure rapid changes in the patient's respiration, at the same time avoiding any appreciable resistance to breathing. As laminar flow depends on fluid viscosity and turbulent flow on density, changes in the character of the gas passing through the pneumotachograph alter its accuracy; for example, changes of temperature or the addition of anaesthetic gases can affect the calibration. To overcome the problem of changes in temperature of the gases passing through them, the heads of some pneumotachographs are maintained at a constant warm temperature by a heating element. This also has the advantage of preventing water vapour in the expired gas

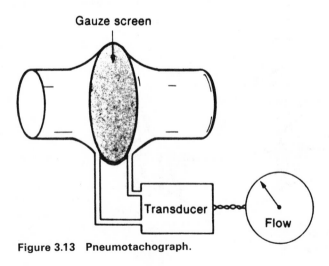

Figure 3.13   Pneumotachograph.

from condensing out on the gauze. In addition to its use to measure flow, the pneumotachograph may be used to record volume by integrating electronically the flow through it (Chapter 25).

Another flowmeter used for calibration or for research purposes is the bubble flowmeter (Fig. 3.14). In this flowmeter a soap solution is used to produce a soap film at the base of a burette. The gas flow is directed up this burette and the rise of the film between two fixed points indicates the flow, the rate of rise being measured by a stop watch. The advantages of this system are that the film is very light and does not obstruct flow and the system is not dependent on the composition of the gases flowing. On the other hand, the bubble flowmeter is suitable only for low flow rates.

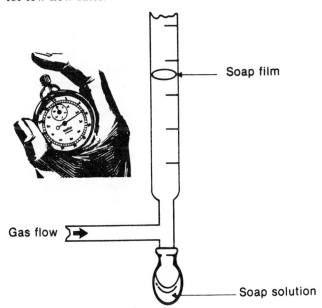

Figure 3.14   Bubble flowmeter.

LIQUID FLOW MEASUREMENT

The measurement of fluid infusion rate is perhaps the most common type of liquid flow measurement in anaesthesia, as the drops passing through the drip chamber of a fluid or blood administration set provide a quick visual indication of flow. In addition, these drops have a nominal volume which permits a measure of the infusion rate to be made. This is obtained by multiplying the drop volume by the rate at which the drops pass through the chamber.

Surface tension is one of the factors determining the final diameter of a drop forming at the end of a tube through which liquid is flowing slowly. As the drop forms, the surface tension on its surface balances the force of gravity on the liquid in the drop and prevents the drop detaching from the end of the tube. As the size increases, the weight of the drop eventually becomes too large to be supported by the surface tension and the drop detaches from the end of the tube. The final volume of the drop depends on the density and surface tension of the liquid, the size and shape of the tube, and the rate at which liquid is flowing through the tube. Density and surface tension depend in turn on the temperature and nature of the solution or of the blood being transfused, so that variations of 20% of the nominal drop volume, and hence of the estimated flow, are possible.

Infusion controllers and pumps are available which determine the infusion rate by counting drops passing through a drip chamber. However, these should not be relied upon to control an infusion with complete accuracy. Infusion sets are available with different calibrations expressed in drops per millilitre, and care should be taken to use the correct set for a particular pump.

The drip counter of an infusion controller is shown clipped onto the drip chamber of an infusion set in Fig. 3.15.

A beam of light or infra-red radiation, produced by a lamp or light-emitting diode, passes through the chamber to strike a photodetector on the other side. The light intensity falling on the photodetector is reduced every time a drop interrupts the beam. Thus, the controller is

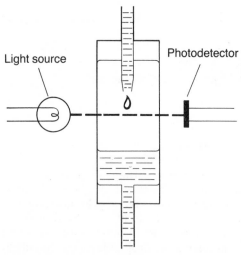

Figure 3.15   Drip counter of an infusion controller.

able to count the drip rate and adjust the flow of liquid to some set value. This adjustment is carried out automatically by the pressure of a bar against a length of delivery tubing clamped in the controller. Movement of the bar flattens the tubing, increasing the resistance to flow and decreasing the rate of infusion.

For reliable operation, drip counters should be positioned carefully so that the light beam is halfway between the drop-forming orifice and the liquid level. The liquid should occupy one-third of the drip chamber, which should be vertical so that the drops interrupt the light beam correctly.

The system described uses the force of gravity to supply the liquid to the drip chamber. Alternatively, an infusion pump may be used to deliver the liquid. In one form of infusion pump a series of bars are moved to pump liquid through the tubing, the drip counter being used to control the flow.

Another form of pump, the volumetric, uses a special syringe cassette with attached tubing which is loaded into a motorised holder. In this instance, the volume of liquid administered and its flow can be accurately known and the problems associated with drip counters are avoided. Ordinary syringes, preloaded with a required drug can also be used in a suitable motorised holder to infuse a small volume of fluid accurately.

High pressures can readily be developed when syringes are used (Chapter 1), and so there is need for particular care when positioning an intravascular cannula for use with an infusion pump.

If it is necessary to check the fluid flow from some apparatus then a measuring cylinder and stop watch may suffice. Variable orifice flowmeters may also be used for liquids as well as gases; for example, such flowmeters may be used to measure flow through a kidney machine or a heart-lung machine. In the case of flows in the circulation, however, more advanced techniques are required and these include dye dilution, isotope dilution and thermal dilution (Chapter 5), ultrasonic measurement (Chapter 13) and the electromagnetic flowmeter (Chapter 14). In addition to these special techniques, the Fick principle is particularly important.

THE FICK PRINCIPLE

The example given in Fig. 3.16 illustrates the principle for the calculation of pulmonary blood flow from the oxygen uptake in the lungs. The patient rebreathes oxygen into a Benedict Roth spirometer through a soda-lime absorber, and the rate of uptake of oxygen is shown by the slope of the tracing to be 250 ml min^{-1}. From a catheter in the right atrium or pulmonary artery a sample of mixed venous blood is obtained and analysed to show an oxygen content of 150 ml litre^{-1} blood. An arterial blood sample shows the arterial blood oxygen concentration to be 200 ml litre^{-1}. Each litre of blood therefore gains an extra 50 ml oxygen in its passage through the lungs; 250 ml oxygen is being taken up in the lungs each minute, therefore $\frac{250}{50}$, i.e. 5 litres blood, must flow through the lungs to carry this volume.

In this example, blood flow has been calculated using the oxygen uptake in the lungs. But, alternatively, the Fick principle could be applied to calculate the blood flow from the pulmonary carbon dioxide excretion and carbon dioxide concentrations in arterial and mixed venous blood samples. The Fick principle can also be applied to measure blood flow to an organ where there is a steady flow of a substance to or from the organ, the concentration of which can be measured.

The general formula is:

$$\text{Blood flow to the organ} = \frac{\text{Rate of arrival or departure of a substance}}{\text{Difference in concentration of the substance in arterial and venous blood}}$$

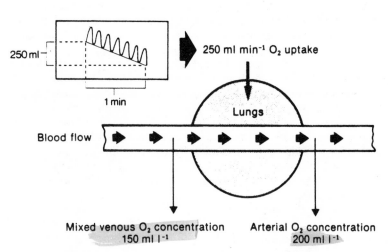

Figure 3.16   Fick principle.

Fig. Cardiac output

4

The Gas Laws

MOLECULAR THEORY

A brief review of molecular theory is given first because it is a help in understanding the gas laws.

All substances are composed of atoms or compounds of atoms, i.e. molecules. In a solid, the atoms or molecules are arranged in a regular formation called a lattice and each molecule in the lattice exerts forces on its neighbours and is continuously in motion, oscillating about a mean position. If heat is added to a solid, each molecule vibrates with a greater amplitude and therefore takes up a greater amount of space. The molecules move further apart and consequently the force exerted by each on its neighbours is reduced. Eventually the forces are not sufficient to hold the molecules in a regular lattice although they still hold them in smaller groupings. The lattice breaks down, the substance melts, and turns into a liquid. In a liquid the molecules still exert some influence on each other and the forces of attraction between them are called Van der Waals' forces. The molecules in a liquid have more vibrational energy than in a solid and each one can move about through the liquid. If heat is added to a liquid, each molecule gains further kinetic energy and eventually some are able to overcome the Van der Waals' forces exerted by their neighbours and are able to move about in space. This state is that of a gas or vapour.

Figure 4.1 illustrates the interface between a liquid and its vapour. Molecules at the surface of the liquid occasionally escape, as shown at 'A' in the diagram. Conversely, molecules of gas moving towards the liquid may transfer into the liquid phase, as shown at 'B'. Eventually, at any one temperature an equilibrium occurs between these two rates of molecular transfer and the vapour above the liquid is said to be saturated.

If the liquid is heated to its boiling point, the energy of the molecules is so great that they all transfer to the gaseous phase. In a gas, the molecules collide with each other and with the walls of a container at frequent intervals. The result of the collisions between the molecules of a gas and the walls of the container is that a force is exerted on the walls, and this force exerted over a certain area is defined as the pressure.

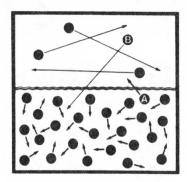

Figure 4.1   Interface between a liquid and a gas.

THE GAS LAWS

Figure 4.2 illustrates a volume of gas within a large syringe. The collisions between the molecules and the walls of the container result in an absolute pressure *P*, which in this case is a typical atmospheric pressure of 100 kPa (1 bar). If this syringe is at a constant temperature, the molecules will have a certain energy of motion and will therefore collide with the walls of the container at a given frequency.

If the temperature is kept constant and the volume of the container is reduced (Fig. 4.3), the molecules will still have the same energy of motion but, as they are in a smaller volume, they will collide with the walls of the container more frequently. The greater the number of collisions with the walls, the greater the pressure of the gas in the

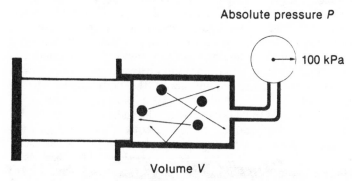

Figure 4.2   Gas in a syringe model to illustrate molecular theory.

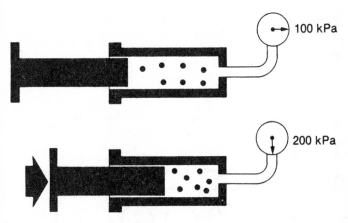

Figure 4.3 Effect of volume change at constant temperature.

container. Halving the volume V of the syringe, as in Fig. 4.3, doubles the absolute pressure P from 100 to 200 kPa.

$$V \propto \frac{1}{P}$$

$$PV = \text{Constant } (k_1)$$

This is the first perfect gas law, Boyle's law.

> *Boyle's law* states that at constant temperature the volume of a given mass of gas varies inversely with the absolute pressure.

Figure 4.4 provides an example of the application of Boyle's law. On the left is shown a typical large oxygen cylinder such as is used in oxygen therapy. How much oxygen will be available at atmospheric pressure? The internal capacity of the cylinder is about 10 litres and when full it has a gauge pressure of 137 bar or 13 700 kPa. If the atmospheric pressure is 100 kPa, the total or absolute pressure of the oxygen will be 13 800 kPa, since absolute pressure is gauge pressure plus atmospheric pressure. The calculations are given on the right of the diagram and the total volume of oxygen works out at 1380 litre. As 10 litre are retained in the 'empty' oxygen cylinder, 1370 litre are available for delivery at atmospheric pressure.

Consider now a gas maintained at constant pressure (Fig. 4.5). In the example shown, this means that the barrel of the syringe is allowed to move freely to maintain the ambient pressure of 100 kPa. If heat is added to the volume of gas, as in the lower diagram, the energy of

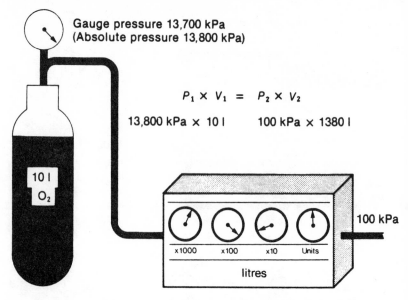

$$P_1 \times V_1 = P_2 \times V_2$$

13,800 kPa × 10 l   100 kPa × 1380 l

Figure 4.4   Use of Boyle's law to calculate the content of an oxygen cylinder.

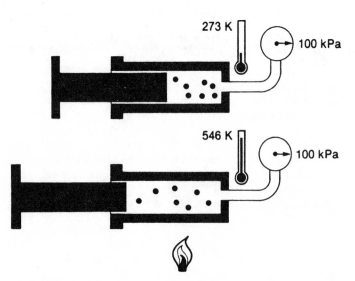

Figure 4.5   Effect of temperature change at constant pressure.

movement of the molecules increases and this results in more collisions with the walls of the container. In order that the frequency of collisions, and thus the pressure, remains constant the volume must increase. Thus, if the absolute temperature T in the syringe is doubled from 273 K to 546 K, it is found that the volume V in the syringe also doubles.

$$V \propto T$$

$$\frac{V}{T} = \text{Constant } (k_2)$$

This is the second perfect gas law, known also as Charles's law or Gay Lussac's law.

> *Charles's law* states that at constant pressure the volume of a given mass of gas varies directly with the absolute temperature.

This law shows that gases expand when they are heated and so become less dense. It means too that warm air tends to rise and this causes convection currents, which are considered in Chapter 9.

The third of the gas laws is illustrated in Fig. 4.6 in which a syringe is maintained at constant volume. If heat is added to this constant-volume container, the molecules of the gas gain kinetic energy and collide with the walls of the container more frequently, thus resulting in an increase

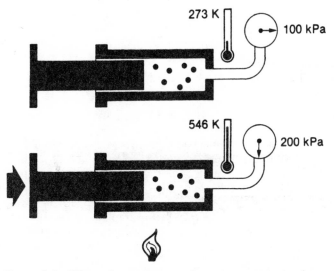

Figure 4.6   Effect of temperature change at constant volume.

in pressure. Consequently, at constant volume, a change in absolute temperature T in a gas produces a change in pressure P.

$$P \propto T$$

$$\frac{P}{T} = \text{Constant } (k_3)$$

The third perfect gas law states that at constant volume the absolute pressure of a given mass of gas varies directly with the absolute temperature.

As an example of this law, consider an oxygen cylinder filled to an absolute pressure of 138 bar at an ambient temperature of 290 K (17 °C). Cylinders are tested to withstand pressures of up to 210 bar. If the cylinder is dropped accidentally into an incinerator at 580 K (307 °C), is there a danger of explosion of the cylinder from the pressure increase? A doubling of the absolute temperature doubles the pressure, thus the pressure in the cylinder increases to over 210 bar. The cylinder is likely to explode even if the weakening of the metal of the cylinder by the heat is ignored.

Another example of this law is the hydrogen thermometer, used as a standard for scientific temperature measurement. When a constant volume of hydrogen is heated, the rise in pressure may be accurately recorded and it gives a measure of the absolute temperature increase.

As volumes of gases are greatly affected by changes of temperature and pressure, it is important to specify the temperature and pressure at which any measurement of volume is made. Moreover, it is often useful to correct results to a standard temperature and pressure. The standard temperature used is 273·15 K (0 °C), and the standard pressure 101·325 kPa (760 mmHg). This standard temperature and pressure is known as s.t.p.

298 Kalvin 72

ADIABATIC CHANGES OF STATE IN A GAS

explosion

The three gas laws describe the behaviour of a gas when one of the three variables, pressure, temperature or volume is constant. For these conditions to apply, heat energy is required to be added to or taken from a gas as the change occurs. The state of a gas can also be altered without allowing the gas to exchange heat energy with its surroundings, and this is called an adiabatic change. An example of such a change occurs when air is compressed in an air supply unit; the temperature of the air rises because of the compression and so a system of cooling is needed (Chapter 21). As a second example, if a gas cylinder connected to an anaesthetic machine or regulator is turned on quickly, the pressure of gas in the connecting pipes and gauges rises rapidly. Thus, the gas is compressed adiabatically and a large temperature rise, with the associated risk of fire, can occur. Alternatively, if a compressed gas expands adiabatically, cooling occurs as in the cryoprobe.

THE CRYOPROBE

The cryoprobe is used for rapid freezing of tissues in the treatment of skin lesions, in gynaecology, and in ophthalmic surgery. When applied to nerves it causes local degeneration of nerve bundles and this results in long-term (3–6 months) local analgesia. This action has been used in the treatment of pain. The cooling of a cryoprobe is the result of an adiabatic process. Gas is allowed to expand rapidly out of a capillary tube and a fall in temperature occurs as a result of the expansion. The cooling effect arises from the fact that energy is required as a gas expands to overcome the Van der Waals' forces of attraction between the molecules of the gas. In a rapid expansion, heat exchange does not take place between the gas and its surroundings, so the energy required comes from the kinetic energy of the gas molecules themselves which results in the gas cooling as it expands.

The elements of a typical cryoprobe are shown in Fig. 4.7. Nitrous oxide or carbon dioxide are suitable gases, and the gas flows from the cylinder through an adjustable pressure regulator which may be used to

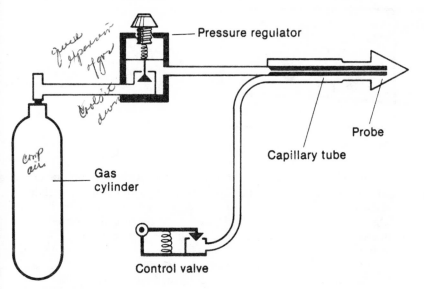

Figure 4.7   Principle of the cryoprobe.

set the cooling rate. The gas flows through a capillary tube in the cryoprobe and expands in the probe tip where a temperature as low as $-70\,^{\circ}C$ may be produced. A foot or hand operated valve may be used to turn off the flow of gas through the probe and hence control the cooling effect.

DALTON'S LAW OF PARTIAL PRESSURES

Figure 4.8 illustrates a mixture of gases in a container. Pressure in the container is related to the frequency of collisions and to the mass and velocity of the molecules of the gas. In a mixture, each type of molecule contributes to the pressure exerted on the walls of the container and because each molecule behaves almost independently of its neighbours the pressure attributable to any one type of molecule is the same whether the other type of molecule is present or not. This phenomenon was first established by John Dalton and is known as Dalton's law of partial pressures.

> *Dalton's law of partial pressures* states that in a mixture of gases the pressure exerted by each gas is the same as that which it would exert if it alone occupied the container.

In anaesthesia the partial pressures of gases in a mixture are often of interest.

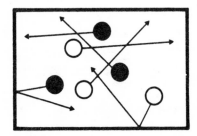

Figure 4.8 Gas mixture in container.

The left of Fig. 4.9 illustrates an Entonox cylinder emptied to an ambient pressure of 100 kPa. The mixture remaining in the cylinder is 50% nitrous oxide, 50% oxygen and so each gas occupies half the cylinder volume. According to Dalton's law, the pressure exerted by the nitrous oxide in the cylinder is the same as it would exert if it alone occupied the container. But if it were to do this, the available space for the nitrous oxide would have increased from half the cylinder to a full cylinder. In other words, it would have doubled its volume. Using Boyle's law it may be calculated that the pressure in the cylinder in this

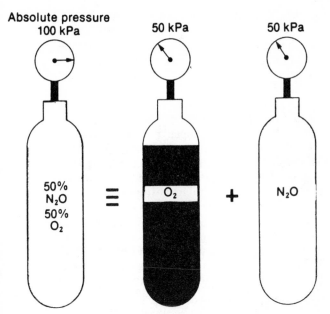

Figure 4.9 Application of Dalton's law in an Entonox cylinder.

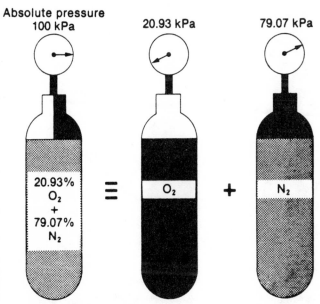

Figure 4.10 Application of Dalton's law in an air cylinder.

case would therefore be halved from 100 kPa to 50 kPa. From these calculations it is seen that the partial pressure of this nitrous oxide is 50 kPa and, similarly, the oxygen pressure is also 50 kPa. So by applying Boyle's law and Dalton's law, the partial pressure of a gas in a mixture is obtained by multiplying the total pressure by the fractional concentration of the gas.

A second example is illustrated by Fig. 4.10, using a cylinder of air at an ambient pressure of 100 kPa. The 20.93% oxygen in the air exerts a pressure of 20.93 kPa and the nitrogen pressure is 79.07 kPa. The calculations can also be made in terms of millimetres of mercury pressure, remembering that an ambient pressure of 100 kPa represents about 750 mmHg. The oxygen partial pressure works out at 157 mmHg, and nitrogen at 593 mmHg.

In the case of humidified gases such as those in the alveoli, the presence of water vapour must be taken into account when calculating partial pressures. Suppose, for example, a meter indicates an end-tidal carbon dioxide concentration of 5.6% measured as a dry gas. Ambient pressure is 101.3 kPa. To find the true pressure of the alveolar carbon dioxide P_{ACO_2}, it is not sufficient to multiply the 5.6% by the ambient pressure as the alveolar gas is fully humidified. From an ambient pressure of 101.3 kPa the water vapour pressure of 6.3 kPa must be subtracted before multiplying by the 5.6%.

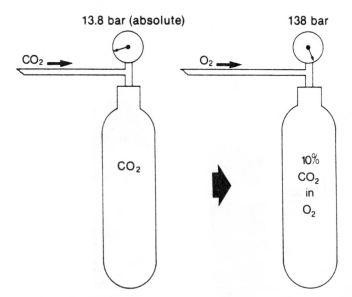

Figure 4.11 Dalton's law in the filling of a cylinder of 10% carbon dioxide in oxygen.

$$P\text{ACO}_2 = (101\cdot3 - 6.3) \times \frac{5\cdot6}{100} \text{ kPa}$$

$$= 5\cdot3 \text{ kPa}$$

The alveolar carbon dioxide pressure is $5\cdot3$ kPa. If the calculation is made in millimetres of mercury, the alveolar carbon dioxide pressure is found to be 40 mmHg. Another example of the effect of humidity on the partial pressure of gases is given in Chapter 19.

Manufacturers make use of Dalton's law when filling cylinders with gas mixtures. Figure 4.11 illustrates the filling of a cylinder to produce a 10% carbon dioxide in oxygen mixture. The cylinder is first filled with carbon dioxide to an absolute pressure of $13\cdot8$ bar. At this pressure carbon dioxide is still gaseous at room temperature. Oxygen is then added to a total absolute pressure of 138 bar. The overall percentage of carbon dioxide is then 10%, the same as the ratio of the pressures.

AVOGADRO'S HYPOTHESIS

Consider the situation illustrated in Fig. 4.12. Two syringes each of volume V are shown containing two different gases, oxygen and hydrogen, maintained at the same temperature. If the appropriate gas molecules are added to each until the pressure in the two syringes is the

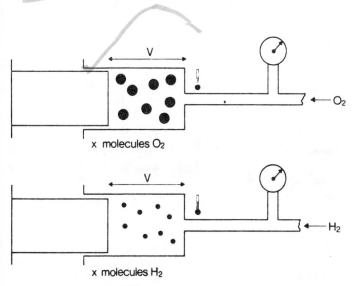

Figure 4.12 Syringes of gas at the same volume, pressure and temperature contain the same number of molecules.

same, it is found that each syringe must contain the same number of molecules. The fact that there are equal numbers of molecules is known as Avogadro's hypothesis.

Avogadro's hypothesis states that equal volumes of gases at the same temperature and pressure contain equal numbers of molecules.

Note that, as the molecular weights of oxygen and hydrogen are different, the masses of the gases in the syringes must be different. Thus, rather than express a quantity of gas in terms of mass, it is more convenient to use a concept related to the number of molecules. This is the mole (Fig. 4.13).

MOLE (mol)

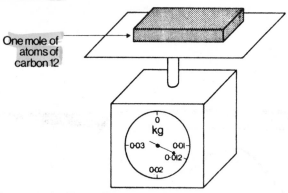

One mole of atoms of carbon 12

Figure 4.13   The mole.

A mole is the quantity of a substance containing the same number of particles as there are atoms in 0·012 kg of carbon 12. It is found that the number of atoms present in 12 g of carbon 12 or the number of particles, for example, molecules, in one mole of any other substance is $6·022 \times 10^{23}$, and this is known as Avogadro's number. It is found that one mole of any gas at standard temperature and pressure occupies 22·4 litre, and so 2 g of hydrogen or 32 g of oxygen or 44 g of carbon dioxide occupy 22·4 litre at s.t.p.

As an example of the mole and Avogadro's hypothesis consider the technique of calibration of a vaporiser illustrated in Fig. 4.14. For convenience, measurements are given as at s.t.p. On the left of the figure a steady stream of oxygen is shown flowing into the vaporiser and completely vaporising the 19·7 g halothane into a volume of 224 litre. What would be the mean percentage concentration of halothane? The molecular weight of halothane is 197, and so 197 g halothane is 1 mol and would occupy 22·4 litre at s.t.p. The 19·7 g halothane in the

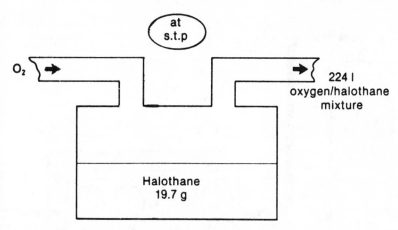

Figure 4.14 Calibration of a vaporiser by the use of Avogadro's hypothesis.

vaporiser is 0·1 mol and would occupy 0·1 × 22·4 litre or 2·24 litre. This halothane, however, has been vaporised into a volume of 224 litre and therefore the concentration of halothane equals 2·24 divided by 224 or 1%.

If measurements were made at temperatures and pressures other than s.t.p., then the appropriate gas laws would need to be applied and suitable corrections made. A very slight error may be introduced because halothane vapour does not obey the gas laws as closely as a gas such as hydrogen, but the results still provide a good approximation.

A similar example is illustrated in Fig. 4.15. A typical full nitrous oxide cylinder contains 3·4 kg of nitrous oxide. The molecular weight of nitrous oxide is 44 and so one mole is 44 g. If the measurements are made at s.t.p., what volume of nitrous oxide is obtained from this cylinder? The calculation is as follows:

44 g (1 mol) nitrous oxide occupies 22·4 litre at s.t.p.

Therefore:

$$3400 \text{ g nitrous oxide occupies } 22\cdot4 \times \frac{3400}{44} \text{ litre} = 1730 \text{ litre}$$

The volume of nitrous oxide at s.t.p. is 1730 litre and this is about 1800 litre at 15 °C.

In practice, the weight of nitrous oxide is used to indicate how full the cylinders are. The weight of the empty nitrous oxide cylinder is known as the tare weight and is always stamped at the top. Consequently, by weighing the cylinder the nitrous oxide content may be calculated.

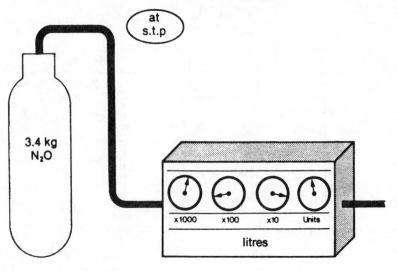

Figure 4.15 Calculation of the volume of nitrous oxide gas available from a cylinder.

UNIVERSAL GAS CONSTANT

The concept of the perfect gas laws can be combined with that of Avogadro's hypothesis and the mole as follows:

$$PV = k_1 \quad \frac{V}{T} = k_2 \quad \frac{P}{T} = k_3$$

Therefore:
$$\frac{PV}{T} = \text{Constant}$$

It is found that PV/T equals a constant for a given quantity of gas and for 1 mole of any gas PV/T equals a unique constant known as the universal gas constant R. The more generally applicable equation with slight rearrangement can be written as $PV = nRT$ where n is the number of moles of the gas and may be greater or less than one.

This formula is applied in anaesthetic practice in the contents gauge of a gas cylinder. The gas cylinder has a fixed volume. Therefore, V in the equation is constant. R is a constant, and if the cylinder is at a fixed temperature, T is constant. Thus, from the formula, P is directly proportional to n, the number of moles. The pressure in the cylinder is therefore directly proportional to the number of moles in the cylinder and so to the amount of gas in the cylinder. The pressure gauge thus acts as a contents gauge provided the cylinder contains a gas.

CRITICAL TEMPERATURE

It has been assumed that all gases obey the gas laws perfectly, but in practice slight deviations occur. Hydrogen obeys the gas laws most closely and for this reason the hydrogen thermometer is used for measurements in the international temperature scale. As a gas cools to near its liquid boiling point, its behaviour deviates from the gas laws.

Figure 4.16 compares the filling of a nitrous oxide cylinder with that of an oxygen cylinder at an ambient temperature of 20 °C. In the top diagram nitrous oxide is being pumped under pressure into the cylinder but, once a certain pressure has been reached, the nitrous oxide liquefies without further increase of pressure until the cylinder is filled to the appropriate level. In the case of oxygen, however, as it is pumped into the cylinder, the pressure gauge indicates accurately how full the cylinder is. No matter how much pressure is applied to the cylinder it is impossible to turn the oxygen into its liquid form at normal room temperature. If, instead, the oxygen cylinder were filled at a very low temperature below −119 °C, then it would be possible to liquefy the oxygen. In other words, it is found that there is a critical temperature for oxygen of −119 °C above which oxygen cannot be liquefied by pressure alone. At or below this temperature, liquefaction under pressure is possible. Each gas has its own critical temperature.

> *Critical temperature* is defined as the temperature above which a substance cannot be liquefied however much pressure is applied.

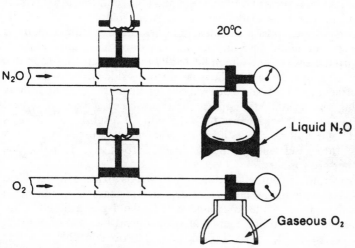

Figure 4.16   Filling of a nitrous oxide cylinder, and an oxygen cylinder.

The critical pressure is the vapour pressure of the substance at its critical temperature. The critical temperature of nitrous oxide is 36·5 °C. Consequently, the nitrous oxide in a cylinder is a gas if the temperature is above 36·5 °C as might occur in the tropics.

Figure 4.17 illustrates a large syringe of nitrous oxide at constant temperature T °C which is above 36·5 °C, and shows graphically the results of compression of the syringe. The line on the graph traces the pressure increase produced as the volume is reduced.

A series of such lines of pressure against volume can be produced at various temperatures and these are known as isotherms.

Figure 4.18 shows the isotherms for nitrous oxide at 40 °C, 36·5 °C, and 20 °C. The top line indicates the changes at 40 °C. As the volume is reduced, moving from right to left on the graph, there is a smooth increase in pressure according to Boyle's law. The curve produced is a rectangular hyperbola. At the critical temperature (36·5 °C) there is a critical pressure of 73 bar at which the nitrous oxide liquefies. One characteristic of a liquid is that it is relatively incompressible and so, when the gas has completely turned to liquid, the slightest decrease in volume is associated with a great increase in pressure. An almost vertical line occurs at this point as shown on the graph. Of greater interest is the bottom isotherm illustrating the results at a room temperature of 20 °C. As the nitrous oxide in the syringe is steadily compressed, at a pressure of 52 bar some of it liquefies as shown in the lower area of the graph. At this point, both liquid and vapour are present and any further decrease in volume causes more vapour to condense and the pressure to remain unaltered. Consequently a horizontal line is present on the graph at 52 bar, and this is the typical pressure in a nitrous oxide cylinder at room temperature, being its saturated vapour pressure. When all the nitrous oxide vapour is condensed into liquid, then any attempt to reduce the volume results in the sharp increase of pressure indicated by the near vertical line in this area of the graph. Note that the word nitrous oxide gas is used for the upper tracing and nitrous oxide vapour for the lower tracing. Strictly speaking the word 'gas' applies to a substance above its critical temperature while 'vapour' is the word used for a substance below its critical temperature. So at normal room temperature, oxygen and nitrogen are gases whereas nitrous oxide, carbon dioxide, halothane and ether are vapours.

Consider the hypothetical case of a nitrous oxide cylinder filled completely with liquid nitrous oxide. Any increase in temperature causes the nitrous oxide to expand but, unlike a gas, a liquid is not compressible. As a result, there would be a great increase of pressure and a considerable risk of explosion. To obviate this risk the manufacturers always ensure that the cylinders are only partially filled in order to leave a volume of nitrous oxide vapour above the liquid.

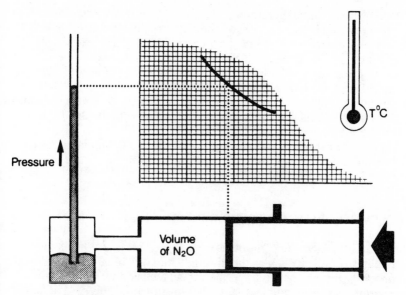

Figure 4.17 Tracing of pressure against volume for a constant quantity of nitrous oxide at a constant temperature.

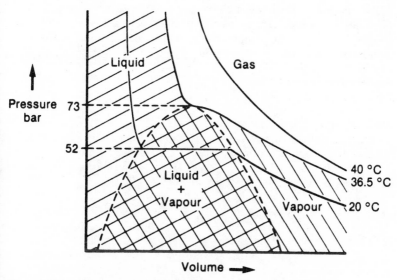

Figure 4.18 Isotherms for nitrous oxide.

Any increase in temperature then causes the liquid to expand, which compresses this vapour and, as the vapour is compressed, some of it condenses, thereby keeping the pressure from rising excessively. In practice, nitrous oxide cylinders are filled according to their weight and for a given size of cylinder there is a fixed quantity of nitrous oxide so that there is no risk of explosion should the cylinder be over-heated.

The term 'filling ratio' is used to describe how much gas is used to fill a cylinder. The filling ratio is the mass of gas in a cylinder divided by the mass of water which would fill the cylinder. As 1 litre water weighs 1 kg, the filling ratio represents the mass of nitrous oxide in kilograms over the internal volume of the cylinder in litres. For nitrous oxide this ratio is 0·65 in the United Kingdom.

PSEUDO-CRITICAL TEMPERATURE

The term 'critical temperature' applies to a single gas. When a mixture of gases is present such as the 'Entonox' mixture of 50% nitrous oxide and 50% oxygen, there is a specific critical temperature at which the gas mixture may separate out into its constituents. This is a different concept to the critical temperature of a single gas, and so the term 'pseudo-critical temperature' is often used for such gas mixtures. In the case of Entonox cylinders it is found that there is a risk of separation if the temperature in the cylinder falls below − 5·5 °C. Such separation is most likely if the cylinder pressure before cooling is 117 bar and is found to be less likely at higher or lower pressures. In Entonox in pipelines, for instance, the pseudo-critical temperature is much lower. It is below − 30 °C at the pipeline pressure of 4·1 bar. Consequently, there is normally no risk of separation of the nitrous oxide mixture in pipelines.

Natural Exponential Functions

THE CONCEPT OF AN EXPONENTIAL PROCESS

A natural exponential function is a special form of non-linear change often encountered in medicine and usually referred to for convenience as an exponential. To simplify the understanding of such a process, the filling and emptying of a bath will be used as an example. Consider first the filling of a bath with vertical sides, as shown in Fig. 5.1.

The height of the water gives a measure of the volume in the bath at any time. A graph of volume against time is linear, as would be expected from the constant flow into the bath.

Figure 5.2 shows the graph describing the emptying of the bath. Initially, emptying is fast because of the high head of water and the high pressure, but as the head of water gets less, the rate at which the bath

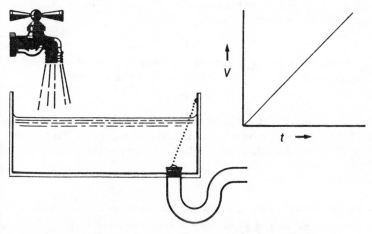

Figure 5.1 The filling of a bath. A constant flow gives a linear relationship of volume *V* to time *t*.

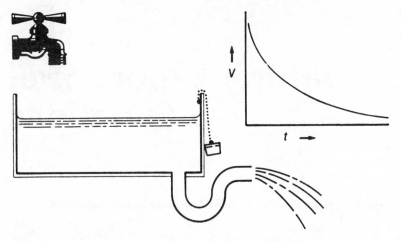

Figure 5.2   The emptying of a bath. A plot of volume V against time t gives a curve.

empties becomes slower. Therefore, a plot of volume against time is a curve, steep initially and gradually getting less steep as the emptying slows, so that in theory the volume never reaches zero. This simple illustration assumes that the water in the reservoir into which the bath is emptying is at the same level as the plughole in the bath. The water in a bath at home would normally be subject to an additional pressure head.

The process may now be analysed in greater detail. If the flow from the bath is assumed to be laminar, then the rate of flow $\dot{V}$ is directly proportional to the pressure head.

$$\dot{V} \propto P$$

On the other hand, P, the pressure head of water driving the flow out, is directly proportional to V, the volume of water in the bath.

$$P \propto V$$

This arises because the bath has vertical sides, so that the height of the water surface above the bottom of the bath is a measure of its volume. The result is that the rate at which the bath empties is proportional to the volume of water in it.

$$\dot{V} \propto V$$

It is the fact that the rate at which the bath empties is proportional to the volume of water in it which makes the process exponential. At any time, the rate of change of the volume is proportional to the volume and this is an example of the definition of an exponential process.

In an exponential process the rate of change of a quantity at any time is proportional to the quantity at that time.

Figure 5.3 gives an example which is more relevant to anaesthetic practice, in which a patient is being ventilated by a constant-flow ventilator. The rate of filling of the lungs is constant, so that a graph of inspired volume against time is linear in the same way that the filling of the bath gives a linear relationship.

Figure 5.4 illustrates expiration. There is a direct analogy to the emptying of the bath and the calculations are similar. The rate of emptying is fast initially as the elastic tissues of the lungs have been stretched during inspiration, but the rate of emptying slows down as the lungs deflate. If quiet respiration is assumed, expiratory flow is predominantly laminar and is proportional to the pressure gradient P.

$$\dot{V} \propto P \left(\text{as laminar flow resistance} = \frac{P}{\dot{V}} \right)$$

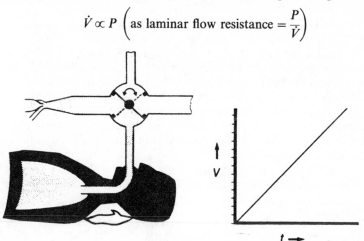

Figure 5.3 Plot of volume against time for inspiration in a patient on a constant-flow generator ventilator.

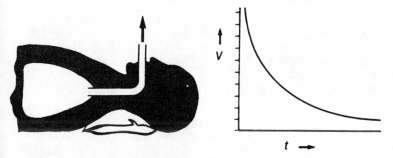

Figure 5.4 Plot of volume against time for expiration.

The pressure difference P in turn is proportional to the volume of the lungs V, if it is accepted that the lungs and chest wall have a constant compliance and if V is defined as the volume of air in the lungs in excess of the functional residual capacity.

$$P \propto V \quad \left(\text{as compliance} = \frac{V}{P} \right)$$

Therefore: $\dot{V} \propto V$

As a result, there is a direct relationship between the rate of emptying of the lungs and the volume of air in them; so the process is exponential.

WASHOUT CURVES

The graphs of such exponential processes are sometimes known as washout curves, and may describe the way in which drugs or anaesthetics are washed out of the tissues by the blood flow. To analyse this process in greater detail another bath is considered in which the water is contaminated by a muddiness which must be washed out (Fig. 5.5). Cleaning out the muddiness could be done by running water in from the tap and running water out of the bath simultaneously at the same flow rate. Again, a simple assumption is necessary; the muddiness must remain uniformly distributed in the bath water as the water flows in and out of the bath.

In Fig. 5.5 a graph of the concentration of mud with time shows an exponential curve. Initially, when the concentration of mud is high, the

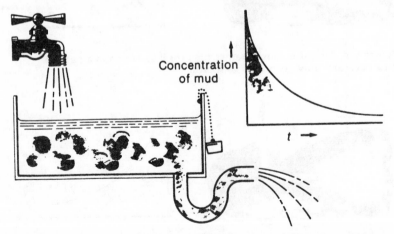

Figure 5.5   Plot of the concentration of mud in a bath against time during its washout from the bath.

mud is washed out at a high rate. Then, as the degree of muddiness falls, the mud is washed out at a slower rate.

The rate at which the mud is removed is the flow of water through the bath multiplied by the concentration of mud in this water.

$$\dot{Q} \propto \dot{V} \times \text{Concentration of mud}$$

where Q = mass of mud present in the bath
$\dot{Q}$ = rate of removal of Q
$\dot{V}$ = rate of flow of water into or out of the bath

but $\dot{V}$ is constant, and therefore:

$$\dot{Q} \propto \text{Concentration of mud}$$

The actual amount of mud Q in the water at any time, however, is proportional to its concentration, as the bath has a constant volume of water in it.

$$Q \propto \text{Concentration of mud}$$

Therefore: $\qquad\qquad \dot{Q} \propto Q$

The process is exponential, as the rate of change of the quantity is proportional to the quantity there at any time.

Washout of this type occurs when indocyanine green is used to measure cardiac output (Fig. 5.6). A known amount of green indicator

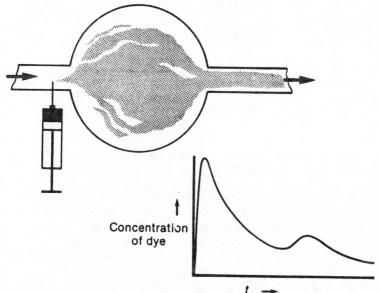

Figure 5.6 Use of a washout curve to measure cardiac output.

dye is injected via a catheter into the right heart. Using the analogy of the previous example, the bath could represent the cardiac cavities and pulmonary circulation and these contain a high concentration of green dye which is steadily washed out by the cardiac output. To sample the dye concentration from the outlet, the concentration of green dye is measured in a peripheral artery using a spectrophotometric technique. If the concentration of dye is plotted against time a curve is obtained as on the right of Fig. 5.6. The initial downstroke of the curve is exponential, but there is an extra hump on the right owing to the recirculation of the dye.

If the cardiac output is high, the washout is much faster as shown on the left of Fig. 5.7; but if the cardiac output is low, the washout of the dye is much slower as illustrated on the right. Recirculation complicates the analysis of the curves, but one technique to overcome this difficulty includes the use of semi-logarithmic paper. This is graph paper on which the vertical scale is logarithmic and the horizontal scale linear (Chapter 25).

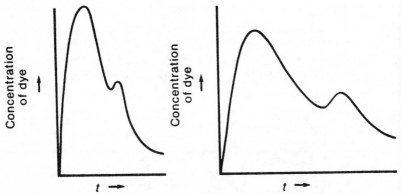

Figure 5.7 Washout curves for a high (left) and low (right) cardiac output.

If the exponential indicator dilution curve is plotted on semi-log paper, the initial downstroke is found to be a straight line (Fig. 5.8). This straight line can be prolonged downwards and from it the stippled area illustrated can be measured. The cardiac output is the amount of dye injected, divided by this area. A logarithmic scale, however, does not have a zero, and in working out the area beneath the line, the continuation of the logarithmic curve that is below the base line of the graph must be taken into account. Microcomputers are often employed to make the calculations and give a readout of cardiac output.

Thermal dilution techniques are often used to measure cardiac output in preference to the dye technique.

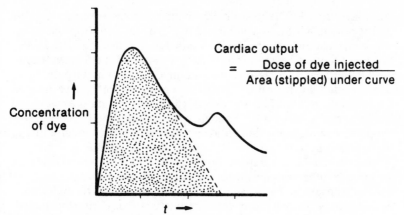

Cardiac output

$$= \frac{\text{Dose of dye injected}}{\text{Area (stippled) under curve}}$$

Concentration of dye

Figure 5.8 Plot of a washout curve on semi-logarithmic graph paper.

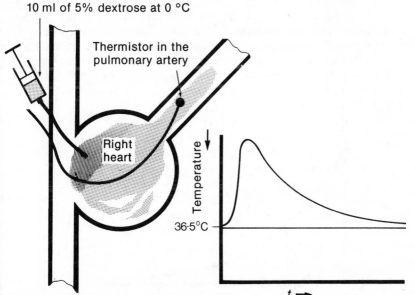

10 ml of 5% dextrose at 0 °C

Thermistor in the pulmonary artery

Right heart

Temperature

36·5°C

Figure 5.9 Thermal dilution technique for cardiac output measurement.

The thermal dilution technique is shown in Fig. 5.9. Two catheters are inserted into the right heart via the internal jugular vein; these catheters may be separate or combined together as a Swan-Ganz catheter. The injecting catheter is inserted to the junction of the superior vena cava and right atrium. The detecting catheter has a thermistor at the end of it and is guided through the right heart into the pulmonary artery. Thermistors are temperature-sensitive devices (discussed in Chapter 9). Ten millilitres of 5% dextrose at approximately 0 °C is injected through the injecting catheter. As it mixes with the warm blood in the right heart the blood cools down, and the temperature of this blood is measured as it passes over the thermistor in the pulmonary artery. A plot of temperature against time gives a washout type curve as shown on the right of Fig. 5.9, and the cardiac output can be calculated from this curve in a similar manner to the dye dilution technique. The temperature scale in the illustration is inverted to show the similarity to the other washout curves. With the thermal dilution technique, however, there is no recirculation to complicate the curve and there is a further advantage in that repeat measurements are easier.

Another example of a washout curve occurs in one of the pulmonary function tests, the nitrogen washout curve. In this case the patient breathes oxygen through a non-rebreathing valve, and the expired gas is analysed.

As shown in Fig. 5.10 there is a steady fall in the nitrogen in the lungs as it is washed out during the oxygen breathing. Normally the end-expiratory nitrogen level should fall below 2.5% after seven minutes. Failure to do so may indicate uneven distribution of gas in the lungs.

Radioactive decay (Chapter 24) is another example of an exponential change. The rate of decay of a radioactive isotope is proportional to the quantity of isotope remaining, and so it is easy to understand why the process is exponential.

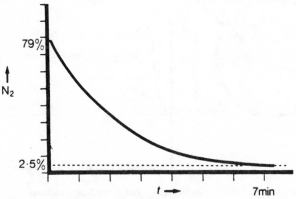

Figure 5.10   Nitrogen washout curve.

DURATION OF EXPONENTIAL PROCESS

In all the examples of exponential processes considered so far, the quantity concerned has been diminishing and these are known as negative exponential functions. Although the quantity is falling, it never actually reaches zero. Consequently, the total length of time taken by the exponential process is infinite and the total time cannot be used to measure the duration of the process. Two alternative systems are used instead: the half-life and the time-constant.

The half-life is illustrated in Fig. 5.11. The half-life is the time taken for the quantity Q to fall to half its initial value. At two half-lives the quantity would of course have fallen to one-quarter of the initial value and to one-eighth after three half-lives and so on.

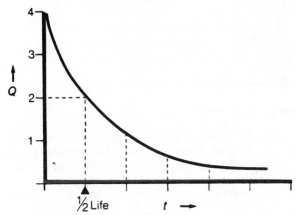

Figure 5.11   Concept of a half-life.

Half-lives are often used to describe the properties of radioactive isotopes. If an isotope has a half-life of one hour, then after one hour the radioactivity is 50% of its initial value, after two hours 25%, after three hours 12·5%, and so on.

As an alternative to the half-life, the rate of an exponential process may be measured by its time-constant.

> *The time-constant* is the time at which the process would have been complete had the initial rate of change continued.

It is given the Greek letter tau (τ) as indicated in Fig. 5.12. After one time-constant the quantity Q has fallen to 37% of its value, the figure of 37% arising for mathematical reasons. Consequently, the time-constant is longer than the half-life.

To help understand the time-constant, consider a definite example such as expiration. In this case a typical time-constant might be 0·3 s, so after 0·3 s (one time-constant) only 37% of the tidal volume is left to be expired, after 0·6 s (two time-constants) 13·5% is left, and after 0·9 s, only 5% of the tidal volume is left to be expired. Thus, expiration is 95% complete after three time-constants.

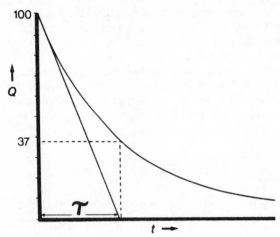

Figure 5.12   Concept of a time-constant.

In a normal person after about one second almost all the tidal volume has been expired and there will be little advantage in prolonging the expiratory period beyond this length of time. In the case of expiration the time-constant can be calculated if the compliance C and resistance R of the lung are known. It is obtained by the product of the two factors:

$$\tau = C \times R$$

If compliance is 0.5 litre kPa^{-1} and the resistance 0.6 kPa s litre^{-1}, then the product is 0.3 s, the figure given above as an example of an expiratory time-constant.

For even distribution of the ventilation in the lungs it is important that different parts of the lungs have the same time-constant, rather than a similar resistance or a similar compliance.

Time-constants have a practical importance for a patient on a ventilator, as the settings for inspiratory and expiratory times are determined to some extent by the time-constant of the patient's lungs.

As illustrated in Fig. 5.13 a high airway resistance leads to a high time-constant and necessitates a prolonged period of expiration to allow expiry of the tidal volume whereas, in the case of a low compliance, the time-constant is reduced and a short period of expiration suffices with a more rapid respiratory rate.

In the case of a washout curve, the time-constant can be shown mathematically to be equal to the volume undergoing washout divided by the flow of the perfusing fluid.

$$\tau = \frac{\text{Volume undergoing washout}}{\text{Flow of perfusing fluid}}$$

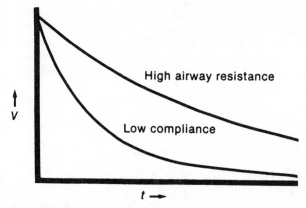

Figure 5.13 Plot of expired air volume against time in a patient with a high airway resistance and in a patient with a low compliance.

The time-constant may be obtained from analysis of a washout curve and if the volume undergoing washout is also known, then the flow of the perfusing fluid may be calculated. Conversely, the volume undergoing washout can be calculated if the time-constant and flow of the perfusing fluid are known. Many methods of measuring blood flow in organs are based on this principle. Radioactive isotopes are often used for such measurements, a radioactive isotope being injected into the organ and the rate at which it is washed out by the blood measured by a scintillation counter (Chapter 24). From this the time-constant is found. As this time-constant is proportional to the volume divided by the flow, if volume is known, flow can be calculated.

The reciprocal of the time-constant is known as the rate-constant, and is sometimes used as an alternative.

THE EXPONENTIAL FORMULA

Now that the exponential type of function has been described, the exponential formula may be considered. The word exponential arises from the exponent or index which is in superscript below:

Fixed exponent   Variable exponent
$$y = x^2 \qquad\qquad y = k^x$$

On the left is an equation or function with a fixed exponent (2 in the example), and on the right is a function with a variable exponent x. The latter is an exponential function.

In medicine, the type of exponential function encountered is normally one in which the rate of change of the quantity is proportional to the quantity at that time as described earlier. This is called a natural exponential function, and the constant k in the above equation has a special value of 2·718. In mathematics this quantity 2·718 is given the symbol e. The exponential formula which describes the emptying of a lung is as follows:

$$V_t = V_0 \, e^{-\frac{t}{CR}}$$

where V_t = volume at t seconds
V_0 = volume at zero seconds
e = 2·718
C = compliance
R = resistance

The volumes V are the volumes of air in the lungs in excess of the functional residual capacity. In this case, e is raised to the power of minus t/CR and the fact that the exponent has a negative value indicates that, as time increases, the value of V_t steadily decreases. Such a negative exponent is characteristic of a negative exponential function.

THE POSITIVE EXPONENTIAL PROCESS

So far, only negative exponentials have been considered but positive exponential processes also exist as shown in Fig. 5.14.

In this case the process is increasing at a faster and faster rate but at any time the rate of change is proportional to the quantity at that time. An example of a positive exponential curve is bacterial growth in the presence of unlimited nutrient, and in some cases the multiplication of

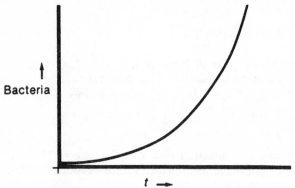

Figure 5.14 Positive exponential curve.

cancer cells may initially follow a similar pattern. Such a curve is sometimes known as an explosive exponential because the rate of the process continually accelerates.

THE BUILD-UP EXPONENTIAL PROCESS

These positive exponential curves must not be confused with another type of exponential, the build-up exponential (Fig. 5.15). This exponential occurs much more often in anaesthesia than the positive form.

In this case the curve represents an inverted negative exponential. The exponential component of the process is the difference between the value achieved and the eventual theoretical final value; this difference is represented by diagonal cross-hatching in Fig. 5.15.

Figure 5.16 shows a bath as an illustration of a build-up exponential

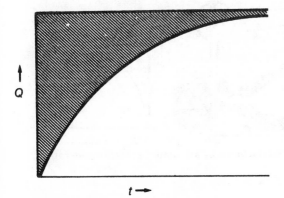

Figure 5.15 Build-up exponential curve.

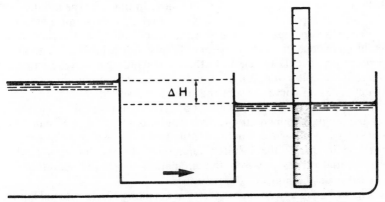

Figure 5.16 A bath analogy for the build-up exponential process.

process. The bath on the right is being filled through its plughole from a constant head of water in a large reservoir on the left. In this case the rate of filling of the bath at any time depends on the difference ΔH, between the height of the water in the bath and the eventual value it will achieve. This height ΔH, represents the pressure required to drive the water through the connecting tube.

As an example of a build-up exponential in anaesthesia, consider the inflation of the lungs with a constant-pressure ventilator (Fig. 5.17).

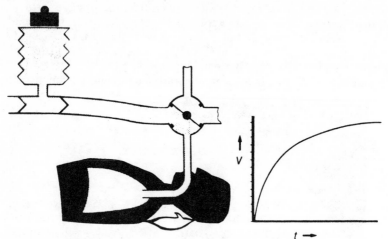

Figure 5.17   Inflation of the lungs by a constant-pressure generator ventilator.

In the ventilator on the left, the weight acting over the area of the bellows gives rise to the constant pressure. In this case the inflation of the lungs is rapid initially but slows as the lungs reach full inflation.

From the analogy with the washout curves discussed earlier these build-up exponentials are sometimes known as 'washin' curves and this concept is particularly helpful to the understanding of the induction of anaesthesia. Such anaesthetic uptake curves are closely analogous to the elimination curves, except that they are inverted.

The uptake of an anaesthetic, however, is not strictly speaking a simple single exponential process, but one consisting of many exponentials in sequence. Firstly, ventilation is one exponential process because it carries the inspired anaesthetic to the lungs; a further exponential process involves the circulation, where the anaesthetic is carried from the lungs to the tissues.

Figure 5.18 illustrates this in terms of the filling of reservoirs of liquid. However, this is a simplified illustration as, in fact, there is not just one

Concentration of
anaesthetic inspired

Concentration of
anaesthetic in
lungs

Concentration of
anaesthetic in
tissues

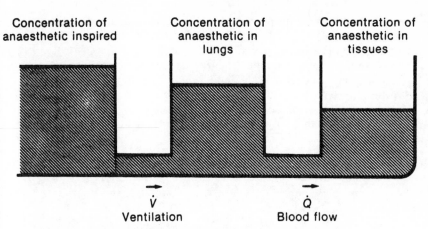

$\dot{V}$
Ventilation

$\dot{Q}$
Blood flow

Figure 5.18 A simplified bath analogy for the uptake of an inspired
anaesthetic agent into the lungs, and tissues.

tissue, represented by the reservoir on the right, but many with different
rates of exponential uptake of the anaesthetic. Elimination of a volatile
anaesthetic is in the reverse direction. It is again a series of exponential
processes.

A complex process, such as the uptake or excretion of anaesthetic,
may be separated mathematically into its component parts. In one
technique, for example, the process may be plotted on semi-logarithmic

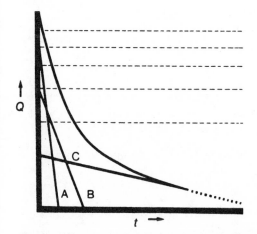

Figure 5.19 A complex process consisting of three exponentials—A, B, and C in the graph—can be analysed into its components.

paper and with a multiple exponential a curve is obtained on the graph instead of the straight line of a simple exponential.

This curve can then be analysed and broken down, as shown in Fig. 5.19. In this simple example the curve is found to be composed of three exponential processes—A, B and C—each giving one straight line. Alternatively, computer techniques are often used to analyse such curves. Other aspects of the uptake and excretion of anaesthetics are considered in Chapter 6.

CHAPTER

6

Solubility

THE CONCEPT OF SOLUBILITY

As explained in Chapter 4, when a liquid is placed in a closed container an equilibrium is eventually established at the surface between the vapour of the liquid and the liquid itself. In this equilibrium state the partial pressure exerted by the vapour is known as the saturated vapour pressure. In practice, there is usually at least one gas present above the surface of the liquid in addition to the vapour.

In Fig, 6.1 molecules of a liquid and its vapour are shown as open circles, and molecules of a gas are shown as black dots. An occasional molecule of gas above the surface passes into solution in the liquid as indicated at Y. In addition, molecules of dissolved gas in the liquid occasionally leave, as indicated at X. Eventually an equilibrium occurs when dissolved gas molecules leave the liquid at the same rate at which others dissolve. In this condition the liquid and the dissolved gas are said to form a saturated solution. As an example, the liquid could be

Gas + vapour

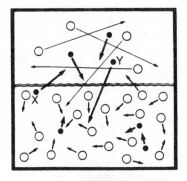

Gas in solution + liquid

Figure 6.1   Molecular equilibrium at the surface of a liquid.

water and the gas molecules could be nitrogen. Suppose the pressure of nitrogen in the enclosed space above the water is doubled, then there will be double the number of molecules in this same given volume, provided that the temperature is constant.

On the left of Fig. 6.2, the gas is at a normal atmospheric pressure of 100 kPa (1 bar) and in equilibrium with the liquid below. On the right, the pressure is doubled to a hyperbaric level of 200 kPa (2 bar); hence there are twice as many molecules of gas above the water resulting in twice as many in solution, as illustrated. For example, the amount of nitrogen dissolved in water is proportional to the pressure of the nitrogen above it. Note, however, that the temperature must be kept constant for this to apply. This principle is known as Henry's law.

> ✓ *Henry's law* states that at a particular temperature the amount of a given gas dissolved in a given liquid is directly proportional to the partial pressure of the gas in equilibrium with the liquid.

The effect of high pressure on the solubility of nitrogen is particularly relevant to deep-sea divers as nitrogen and other gases, if breathed under pressure, pass into solution in the tissues. If a return to atmospheric pressure is made too rapidly, the nitrogen comes out of solution as small bubbles in the joints and elsewhere, giving rise to the condition known as decompression sickness or the 'bends'. Similar

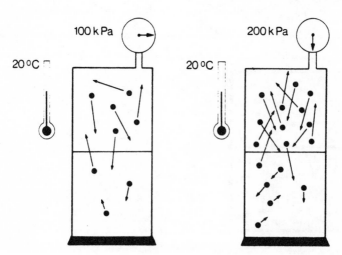

Figure 6.2   Effect of pressure on the amount of gas dissolved in a liquid.

problems apply to workers in hyperbaric chambers, where again a decompression that is too rapid can lead to the 'bends'.

Henry's law applies only if the temperature is constant. Consider now what happens if the temperature is altered.

Figure 6.3 shows two measuring cylinders, each containing 1 litre of water and exposed in an atmosphere of nitrogen. The left cylinder is at a room temperature of 20 °C, the right one at body temperature, 37 °C. The pressure of the nitrogen in both cases is 100 kPa but the volume of nitrogen that is dissolved in the litre of water in the cylinder at room temperature is 0·016 litre, while at body temperature it is only 0·014 litre. It is a general property that as a liquid is warmed, less gas dissolves in it. In a kettle about to boil the simmering which occurs is due to the release of bubbles of air which come out of solution as the water is heated. Similarly, in the operating theatre bubbles of air may form in saline in an infusion line which has passed through a blood-warming coil.

Consider now what happens if a gas other than nitrogen is used. The measuring cylinder on the left of Fig. 6.4 contains 1 litre water in an atmosphere of nitrogen and on the right is a similar cylinder in an atmosphere of nitrous oxide. At equilibrium there is considerably more nitrous oxide dissolved in the litre of water than in the case of nitrogen, 0·39 litre of nitrous oxide compared with 0·014 litre of nitrogen. So different gases are found to have different solubilities.

Finally, the liquid too must be specified when considering solubility.

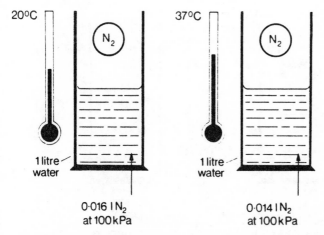

Figure 6.3 Effect of temperature on the amount of gas dissolved in a liquid.

As shown in Fig. 6.5, 0·39 litre nitrous oxide dissolves in 1 litre water at body temperature, whereas 1 litre blood dissolves 0·47 litre of nitrous oxide under the same conditions. To summarise, the solubility of a gas depends on the partial pressure, the temperature, the gas and the liquid concerned.

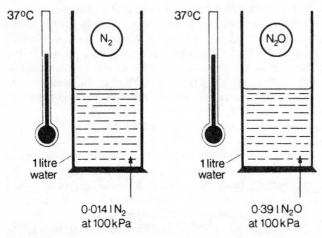

Figure 6.4   Effect of the gas concerned on solubility.

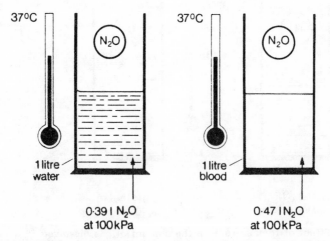

Figure 6.5   Effect of the liquid concerned on the solubility of a gas.

Vol of gas dissolve in a liq. 1 liter of water

SOLUBILITY COEFFICIENTS

proportion of gas that dissolves in a liquid

The solubility of solids is described in terms of a given number of moles or of millimoles of the solute which dissolve in 1 litre of the solvent, but for gases it is, in practice, much easier to express a quantity in terms of its volume as in the previous examples. The volume of a gas, however, only has an accurate meaning if the pressure and temperature are specified, and for scientific work the volume of a gas is normally corrected to s.t.p. Physicists therefore use a solubility coefficient, known as the Bunsen solubility coefficient, which utilises such a correction.

> *The Bunsen solubility coefficient* is the volume of gas, corrected to s.t.p., which dissolves in one unit volume of the liquid at the temperature concerned, where the partial pressure of the gas above the liquid is one standard atmosphere pressure.

The Bunsen coefficient is the one usually quoted in scientific tables and textbooks but in anaesthetic practice another solubility coefficient is preferred. This is known as the Ostwald solubility coefficient.

> *The Ostwald solubility coefficient* is the volume of gas which dissolves in one unit volume of the liquid at the temperature concerned.

The volume of gas is not corrected to standard temperature and pressure, in this case but, instead, is measured at the ambient temperature and pressure. The practical advantage of the Ostwald solubility coefficient is that it is independent of pressure.

Figure 6.6 illustrates this independence of pressure. On the left, there is 100 kPa pressure (1 bar) of nitrogen above the surface of 1 litre of water. A volume of 0·016 litre nitrogen at that pressure dissolves in the water. On the right, there is a pressure of 200 kPa (2 bar) of nitrogen above the water. Consider what happens. According to Henry's law, the quantity of the gas dissolving is proportional to the pressure, so there is twice as much nitrogen dissolved in the water on the right, i.e. 0·032 litre at 100 kPa pressure, But if this volume were measured at the ambient pressure of 2 bar it would be halved, i.e. to 0·016 litre (Boyle's law). Consequently, the volume of nitrogen dissolved measured at the ambient pressure remains constant even though the partial pressure of the nitrogen is doubled and the number of nitrogen molecules in solution is also doubled. The Ostwald solubility coefficient does not therefore need to be defined in terms of pressure because pressure does not modify the volume which dissolves, provided that the volume is measured at the ambient pressure.

Solubility of O_2 1 liter blood 200 ml O_2
.2 = 20%

20cc O_2/ 100cc Bl

How much O_2 in plasma   3 ml/l.

$\frac{200}{3}$

Hgb # solubility of O_2 in blood by 70 times

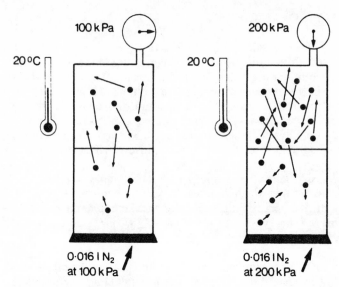

Figure 6.6 The illustration shows that the Ostwald solubility coefficient is independent of pressure.

Blood Gas partition Coefficient
For every cc of gas that exists above
23/1   blood, 23 cc exists in blood.

✓ THE PARTITION COEFFICIENT

The partition coefficient is similar to the Ostwald solubility coefficient.

To understand the term partition coefficient consider the example of nitrous oxide and blood shown in Fig. 6.7. There is a unit volume of 1 litre nitrous oxide above the same unit volume of 1 litre blood which contains dissolved nitrous oxide in equilibrium at 37 °C. The volume of nitrous oxide dissolved in these circumstances is 0·47 litre and the ratio of 0·47 to 1 is the blood–gas partition coefficient for nitrous oxide.

> *The partition coefficient* is defined as the ratio of the amount of substance present in one phase compared with another, the two phases being of equal volume and in equilibrium.

As with the Ostwald coefficient, the temperature should be specified and, in addition, the phases concerned must be stated. In Fig. 6.7 the temperature is 37 °C and the phases are blood and gas and the nitrous oxide blood–gas partition coefficient is 0·47, the same as the Ostwald solubility coefficient. Notice, however, the relative order of the phases of the partition coefficient must be clearly identified. For instance, if the nitrous oxide gas–blood coefficient were specified instead of the blood–gas coefficient, then the value would be 2·1 instead of 0·47.

more gas dissolved in a liquid that exists above the liquid.

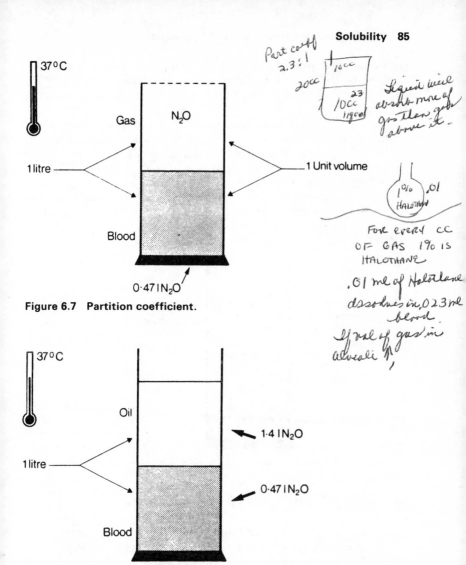

Figure 6.7 Partition coefficient.

Figure 6.8 Application of the partition coefficient to two liquids.

Another difference between the partition coefficient and the Ostwald solubility coefficient is that the former can be applied to two liquids.

Figure 6.8 shows an illustration with two liquids, blood and oil. At equilibrium, the ratio of the amount of nitrous oxide present in a unit volume of blood to that in a unit volume of oil is the blood–oil partition coefficient. In this case, 1·4 litre nitrous oxide dissolves in 1 litre oil, compared with 0·47 litre in 1 litre blood. Thus, the blood–oil partition coefficient for nitrous oxide is 0·47/1·4, or 0·33. In the case of partition coefficients the nitrous oxide or other dissolved gas must be in equilibrium in the two phases, i.e. the tension must be the same in the two phases.

The word tension is often used in place of partial pressure for gases in solution and Fig. 6.9 shows the pressure or partial pressure of nitrous oxide in the gaseous phase above blood as 50 kPa. When in equilibrium, the tension of nitrous oxide in the blood is also 50 kPa. The tension of a gas in solution is the partial pressure of the gas which would be in equilibrium with it. The word tension is also used in physics in another sense when referring to forces (Chapter 1).

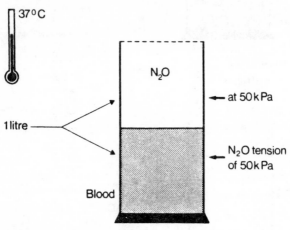

Figure 6.9   Tension of a gas.

SOLUBILITY AND UPTAKE OF ANAESTHETICS

Some clinical applications of solubility may now be considered.

Figure 6.10 shows three measuring cylinders, all containing 1 litre blood at a temperature of 37 °C and exposed to an atmosphere of nitrous oxide, halothane, and ether, respectively. There are many more molecules of ether present in the blood because it has the highest Ostwald solubility coefficient (12), higher than halothane (2·3) and nitrous oxide (0·47).

How does the increased solubility of the ether affect the rate at which it can be transported away from the lungs in the blood? Because it is so very much more soluble, ether is carried away from the lungs more rapidly than halothane or nitrous oxide and hence the concentration of ether in the alveolar air builds up more slowly than that of a less soluble anaesthetic.

If cardiac output ↓ - Go to sleep faster.
⍦ less movement of inhalant
C·O↑ - takes more gas to put to sleep   **Solubility 87**

As explained in Chapter 5, the uptake of an anaesthetic can be represented diagrammatically by a set of reservoirs, and Fig. 6.11 shows such a system to illustrate the effect of solubility. The reservoirs on the left represent the inspired anaesthetic concentration. The transport of the anaesthetic to the lungs by the ventilation is represented by the pipe joining it to the second reservoir and the ventilation is indicated by $\dot{V}$. The second or central reservoir represents the lungs, i.e. the concentration of anaesthetic in the alveolar air. The removal of the anaesthetic by the pulmonary blood flow from this reservoir is denoted

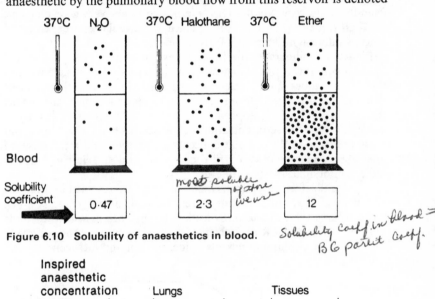

Figure 6.10   Solubility of anaesthetics in blood.

most soluble of three we use

Solubility Coeff in blood = BG partit. Coeff.

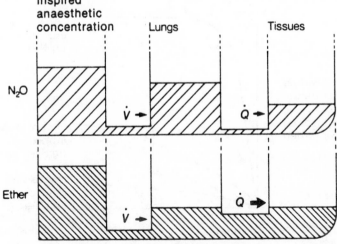

Figure 6.11   Analogy of the uptake of an anaesthetic to the filling of a series of reservoirs.

The more soluble gas is in blood, more is taken out of alveolus.

by the next pipe and the rate of transport by the cardiac output is shown as $\dot{Q}$. The right reservoir represents the tissues.

The upper diagram illustrates the uptake of nitrous oxide. The central reservoir represents the alveolar concentration of the anaesthetic, and its narrow outlet pipe indicates that the anaesthetic can be carried away only relatively slowly by the cardiac output because the solubility coefficient of nitrous oxide is only 0·47.

The lower diagram shows what happenes when ether is used. The high solubility of ether in the blood means that it can be carried away from the lungs very rapidly. This is indicated by the very wide pipe and the high value of $\dot{Q}$ and, in consequence, the alveolar concentration of ether is only slightly above the tissue level.

Figure 6.12 illustrates the build-up exponential curves produced if alveolar concentrations of anaesthetics are expressed as a percentage of their final values and plotted against time. It is seen that soluble anaesthetics such as ether are very slow to approach their final value, whereas equilibrium is approached in 10 to 15 minutes with nitrous oxide. Because the concentrations of anaesthetics in the blood and brain are close to the alveolar concentrations, there is a rapid onset of anaesthesia in the case of nitrous oxide and a relatively slower induction of anaesthesia with ether.

In these illustrations, nitrous oxide has been represented as an agent of low solubility, but solubility is relative and nitrous oxide is much more soluble than nitrogen. Consider the instance in which a simple pattern of a Benedict Roth spirometer is used to measure the volume of a gas mixture containing nitrous oxide, oxygen and nitrogen.

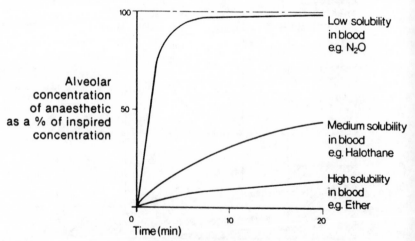

Figure 6.12 Graph of alveolar concentrations of anaesthetics against time during anaesthetic uptake.

As shown in Fig. 6.13, more nitrous oxide passes into solution than oxygen or nitrogen and this could lead to a slight error, although with the small volume of water in the water seal of modern spirometers this error is minimal. A greater loss of nitrous oxide into solution would have an effect on the concentrations of the other gases, oxygen and nitrogen, in the bell of the spirometer—resulting in a rise in the concentrations of oxygen and nitrogen. A similar effect occurs in the alveoli during induction of anaesthesia. During the course of an inspiration of a gas mixture containing nitrous oxide, the nitrous oxide is absorbed into the bloodstream faster than the oxygen or nitrogen and so at peak inspiration, when the pressure in the alveoli has equalised with the ambient pressure, there is a net surplus of oxygen and nitrogen molecules. This phenomenon is called the second-gas effect and is important when considering the uptake of anaesthetics.

A similar phenomenon occurs during the excretion of nitrous oxide at the end of anaesthesia when nitrous oxide leaving the blood and entering the alveoli dilutes the gases already present. The fall of oxygen concentration thus produced is known as diffusion hypoxia.

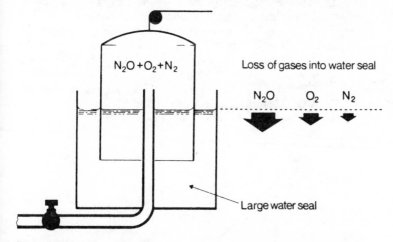

Figure 6.13 Potential error in volume measurement due to gas solubility in a water seal.

OIL SOLUBILITY AND ANAESTHETIC POTENCY

Fat is an important constituent of many body tissues, being present in cell membranes and neurones in particular. Fat and oil are very similar and, because it is easier to measure solubilities in oil, it is used for measurements.

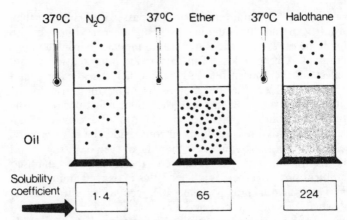

Figure 6.14   Solubilities of anaesthetics in oil.

Figure 6.14 illustrates the relative solubilities of nitrous oxide, ether and halothane in oil, the greater the solubility the greater the number of dots representing the molecules of anaesthetic agent in the oil. The solubilities in oil are in a different order from those in blood (Fig. 6.10). Although nitrous oxide is still the least soluble, the agent with the greatest solubility is now halothane as shown on the right, ether being midway between nitrous oxide and halothane.

Agents that have the highest oil solubility are found to have the greatest potency as anaesthetics and this is the basis of the Meyer-Overton theory of anaesthesia, illustrated in a modified form in Fig. 6.15.

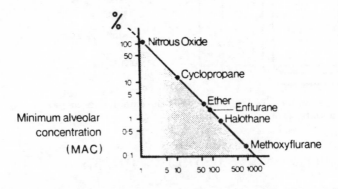

Figure 6.15   Graph of minimum alveolar concentration (MAC) of anaesthetics and their oil solubility coefficients.

The vertical axis of the graph is the MAC (minimum alveolar concentration) value for the anaesthetic; the MAC value being the point at which 50% of patients cease to move in response to a stimulus. On the horizontal axis are the oil solubility coefficients of the anaesthetics and in order to cover the range of anaesthetics logarithmic scales have been used. When plotted in this way the anaesthetics are found to lie on a straight line and from the slope of the line it is seen that an anaesthetic with a high solubility in oil has a low MAC value. In other words, an anaesthetic with a high oil solubility is effective at a low alveolar concentration and has a high potency.

An alternative way of expressing this principle is to say that the MAC value for any anaesthetic is related to the number of molecules of anaesthetic dissolved in oil, and this applies to all volatile anaesthetics. The MAC value of over 100% for nitrous oxide shown in the graph is, of course, a theoretical one based on the partial pressure of nitrous oxide when used for anaesthesia in a pressure chamber.

To explain the inter-relation of the potency of anaesthetics and their solubility in oil, the high solubility of potent anaesthetics is thought to be associated with a mild attraction of the anaesthetic molecules to those of fatty-acid chains. Anaesthetics may act at some point in the brain by interfering with the molecular configuration of the long fatty-acid chains or similar molecular chains at a critical point within the neurones or their synapses. The attachment of the anaesthetic molecules to the chain molecules is relatively loose and readily reversible with Van der Waals' type bonds.

Anaesthetic molecules are also attracted to the long carbon molecule chains present in rubber. The solubility of anaesthetics in rubber results in changes in the physical properties of the rubber and this has practical importance in anaesthesia. Considerable quantities of anaesthetics are absorbed into the rubber tubing used in the breathing systems, and the more soluble the anaesthetic in the rubber the greater the volume which is absorbed in this way. For example, potent anaesthetics such as methoxyflurane and halothane can be absorbed in considerable quantities and trace amounts of these anaesthetics may later be released and administered to patients even if the anaesthetic vaporisers are not turned on.

when MAC highest in life (6mo).
what affect MAC.
age

CHAPTER
7

Diffusion and Osmosis

Diffusion is the process by which the molecules of a substance transfer through a layer or area such as the surface of a solution. In the lungs, gases diffuse across a gas–liquid barrier in this way, but diffusion may also occur at other sites, shown diagrammatically in Fig. 7.1.

The diffusion of gases in the body is illustrated by compartments in the form of layers in a bottle. At the top, gas in the alveoli of the lungs is represented and below it the alveolar–capillary membrane where the gas diffuses into the liquid phase, i.e. the blood.

There are then further membranes at which diffusion occurs, between the blood and the extracellular fluid, and between the extracellular fluid and the cell, allowing the gas to pass from one compartment of the body to the next.

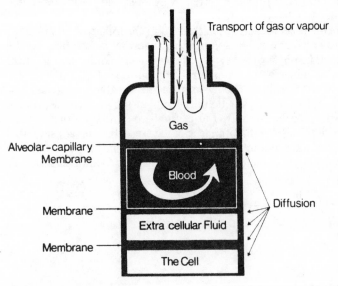

Figure 7.1 Simple model to illustrate the diffusion of a gas or vapour in the body.

Diffusion can still take place without a membrane or a gas–liquid barrier. For example, if gas escapes from a broken gas pipe, the gas spreads by diffusion even after the gas tap has been turned off.

FACTORS AFFECTING THE RATE OF DIFFUSION

Figure 7.2 provides an example in which two different concentrations of a substance or gas diffuse into an area of lower concentration. This could represent the gas from the broken pipe in the example above, or an anaesthetic diffusing across a membrane, or a gas diffusing into a liquid. On the right, the concentration is double that on the left and it is found that the rate of diffusion through the surface or membrane is doubled. If there is a gas concentration below the membrane, an allowance for this must be made and the rate of diffusion would depend on the concentration gradient across the membrane.

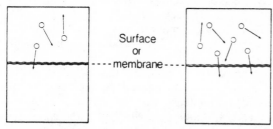

Figure 7.2 Effect of concentration on the rate of diffusion.

The law defining this relationship is Fick's law of diffusion.

> *Fick's law* states that the rate of diffusion of a substance across unit area (such as a surface or membrane) is proportional to the concentration gradient.

Fick's law applies only in a single homogenous phase and a slight modification is needed if gases are transferring from one phase into another—as in the case of gases passing into solution.

The modified form of the law is that the rate of diffusion of a gas across a membrane (or surface area) is proportional to the tension gradient.

The tension of a gas in solution describes the partial pressure of the gas which would be in equilibrium with it, as explained in Chapter 6. For convenience the word tension is often used instead of partial pressure for the gas in the gaseous phase. This modified form of Fick's law is more relevant in anaesthesia than the original form because it applies universally. For example, it is relevant to clinical situations in which a membrane or interface exists between gases, or between liquids, or between gas and liquid.

These laws apply only to a given gas and a given membrane because different gases have different rates of diffusion, and membranes also vary in their diffusing capacity.

Fick's law of diffusion may now be considered in the context of the transport to the cell of a gas or vapour. At the top of Fig. 7.1 inspired gas is illustrated being transported by ventilation and at this point the rate at which it is transported would be the flow multiplied by the concentration. Once the gas reaches the alveolar–capillary membrane, diffusion takes place and here the tension gradient is important. In the circulation, again there is mass transport, the rate being flow times concentration, but from the blood to the cell there are two membranes with the extracellular fluid compartment between them. Diffusion takes place at these points and so the rate is affected by the tissue tension gradients. In addition to diffusion at these points, however, the carriage of the gases may also be helped by fluid movements in the extracellular fluid caused by osmosis (discussed later).

The same patterns of diffusion and mass transport apply in reverse for a substance such as carbon dioxide which is produced at the cell and excreted by the ventilation.

Consider now diffusion across the alveolar-capillary membrane (Fig. 7.3). Here the gas passes into the liquid phase and, once in solution, diffusion of the gas depends on the tension or concentration gradients in the liquid itself. The greater the solubility of the gas in the liquid, the greater will be the concentration gradient between the surface layer of the liquid and its deeper layers; hence a more soluble gas diffuses into the deeper layers of the liquid more rapidly. This maintains a tension gradient between the gas and the surface layer of the liquid and

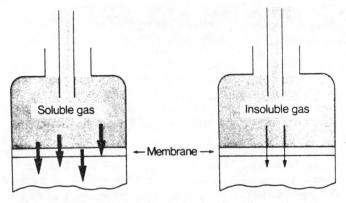

Figure 7.3 Diffusion across the alveolar–capillary membrane.

increases the rate of diffusion of the gas into the liquid. The diffusion of a gas across a membrane into or out of a liquid is thus proportional to its solubility in the liquid.

Figure 7.4 illustrates this for carbon dioxide and oxygen. Because carbon dioxide is more soluble in tissue fluid than oxygen, it diffuses more rapidly. Below the diagrams, graphs illustrate the rate of diffusion across the alveolar–capillary membrane for carbon dioxide and oxygen. The blood normally takes 0·75 s to pass through the capillary, but carbon dioxide equilibrates virtually instantaneously in less than 0·1 s. Oxygen equilibrium is normally complete by the time the blood has passed half way down the capillary, the longer time taken for the oxygen being due to two factors—its slower diffusion compared with carbon dioxide and the removal of the oxygen molecules from the plasma as they combine with haemoglobin in the red cells. If diffusing capacity is limited, hypoxia is thus more likely to occur than hypercapnia. Nevertheless, there is still a reserve of time for oxygen to reach equilibrium before the blood leaves the capillary and mild limitation of diffusion does not normally produce symptoms.

The diffusion of volatile anaesthetics resembles carbon dioxide, equilibrium being complete within a small fraction of a second and so diffusion is ignored in Chapter 6 when considering the uptake of anaesthetics.

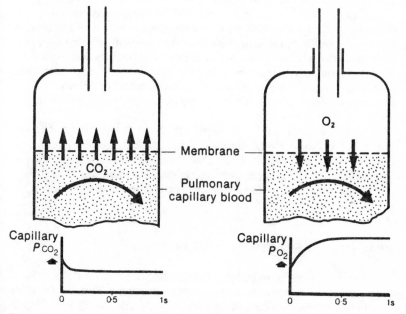

Figure 7.4   Diffusion of carbon dioxide and of oxygen.

PULMONARY DIFFUSING CAPACITY AND CARBON MONOXIDE TRANSFER TEST

Diffusion across the alveolar–capillary membrane is altered by certain diseases and can be measured by a method using carbon monoxide. This measurement, known as the pulmonary diffusing capacity or transfer test, is the rate of carbon monoxide uptake, standardised according to the tension gradient. Because carbon monoxide is very toxic, it can be used only in trace levels, but a patient can safely inhale a trace level of carbon monoxide of 0·1% into his alveoli for a short period. As haemoglobin has a high affinity for carbon monoxide, the carbon monoxide molecules disappear rapidly into the cells in the bloodstream and are carried away. Consequently, it would take a very long time for the plasma in the pulmonary capillaries to equilibrate with the alveolar gas and produce a significant carbon monoxide tension. The carbon monoxide tension in the pulmonary capillaries during this test can therefore be regarded as about zero. The rate at which the carbon monoxide leaves the alveoli is therefore dependent on the rate of diffusion through the membrane and not on the pulmonary blood flow. In this way the rate of uptake of trace concentrations of carbon monoxide is used to measure diffusing capacity.

As lungs have a large reserve of diffusing capacity, small reductions are of little importance in patients. For example, a pneumonectomy halves the available membrane area and so approximately halves the diffusing capacity, but even a pneumonectomy does not normally give rise to any serious problems from this reduction of diffusing capacity. Certain diseases however, such as sarcoidosis, asbestosis and sometimes systemic lupus erythematosis, affect the alveolar–capillary membrane itself and can give rise to a severe reduction of diffusing capacity, and so the measurement of the diffusing capacity may be of value in the diagnosis of these diseases.

As illustrated in Fig. 7.1, there are further membrane barriers between the blood in the capillaries and the extracellular fluid, and between the extracellular fluid and the cells. Diffusing capacity is altered if oedema is present, because oedema fluid impairs the diffusion of gases or other agents to the cells by increasing the distance from the capillary to the cell surface.

PRACTICAL ASPECTS OF DIFFUSION

Consider now some practical points regarding diffusion and anaesthetic apparatus. Anaesthetic systems deliver halothane vapour through corrugated rubber tubing and, as halothane is very soluble in rubber it diffuses readily into and through the rubber of the tube. The rubber tube thus provides a possible source from which halothane may diffuse into the theatre atmosphere, in addition to the more direct route from release of halothane vapour through the expiratory valve. As mentioned in Chapter 6, halothane may remain stored in the rubber due to its high solubility, and trace levels may be released later.

As another example of the effect of tension gradients on diffusion, consider a patient with decompression sickness in whom bubbles of nitrogen form in the tissues.

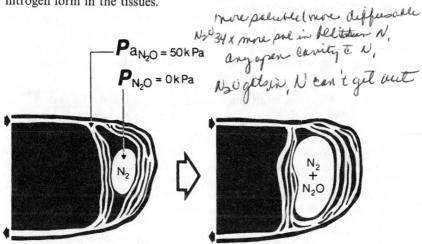

$P_{a_{N_2O}} = 50\,\text{kPa}$

$P_{N_2O} = 0\,\text{kPa}$

[handwritten annotations:] more soluble/more diffusable
N_2O 34 x more sol in blood than N_2
any open cavity c N_2
N_2O gets in, N can't get out

Figure 7.5 Effect of nitrous oxide on the size of a bubble of nitrogen in the tissues.

On the left of Fig. 7.5 is a bubble of nitrogen in the tissues within a capillary loop. If the patient then receives anaesthesia with 50% nitrous oxide, a tension gradient develops as shown and the arterial tension of nitrous oxide could be 50 kPa compared with zero tension within the bubble. Nitrous oxide diffuses into the bubble thus enlarging it, as shown on the right, and aggravates the condition. This problem is not limited to patients with decompression sickness but applies equally to those with bubbles of air at other sites such as pneumothorax. Even air in the middle ear expands and this may give rise to problems in grafts in that region. In Fig. 7.5 there is a reverse gradient for nitrogen, but the nitrogen is not removed from the bubble at the same rate that nitrous oxide diffuses into it because the lower solubility of nitrogen restricts the rate of its diffusion and the rate at which the capillary bloodstream removes it. Another example of nitrous oxide diffusion arises routinely during the course of anaesthesia, when the gas diffuses into the cuff of an endotracheal tube so increasing the pressure on the tracheal mucosa.

EFFECT OF MOLECULAR SIZE

Consider now the effect of varying the size of the molecules of the substance which is diffusing.

In Fig. 7.6 there is a broken ether bottle on the floor of a room in which ventilation is negligible; the ether vapour has formed a smooth layer over the floor, ether being more dense than air. On the right, a gas pipe has broken and has released hydrogen into the upper part of the room. Because small molecules move faster than large molecules at the same temperature, the hydrogen diffuses more rapidly than the ether and the ether vapour tends to remain longer as a layer.

The relationship of the molecular size to the rate of diffusion is defined by Graham's law of diffusion.

✓ *Graham's law* states that the rate of diffusion of a gas is inversely proportional to the square root of its molecular weight.

When dealing with membranes, the law applies only in certain cases and it does not apply accurately when more complex biological membranes are involved. Although Graham's law applies to gases, the concept that the rate of diffusion is related to molecular size may also be applied to liquids.

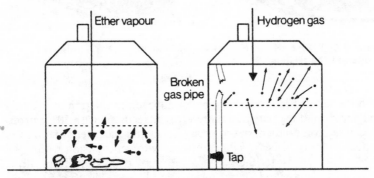

Figure 7.6   Effect of molecular size on diffusion rate.

On the left of Fig. 7.7 bupivacaine is shown, injected underneath a layer of serum so that it forms a smooth layer. The molecular movement in liquids is much slower and more limited than in the case of gases, and this is illustrated in the diagram by the shorter arrows. There is a very slow rate of diffusion in liquids such as serum compared with the rate of diffusion in vapours or gases. When injecting local anaesthetics into the tissues it is important to inject the anaesthetic

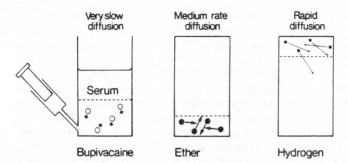

Figure 7.7 Diffusion in a liquid, e.g. serum, is slower than for a vapour or gas.

close to the nerves concerned, because diffusion allows only limited penetration of the local anaesthetic into the tissues. In the case of cerebrospinal fluid there are also convection currents and gravitational effects depending on the density of the local anaesthetic. All these effects help the local anaesthetic to reach the appropriate nerve roots.

The main points regarding diffusion may be summarised as follows. Diffusion is proportional to the tension gradient and, because solubility has a pronounced effect on tension gradients, this in turn affects the rate of diffusion. Diffusion also depends on the membrane concerned, its area, thickness and its constituents. Finally, diffusion is affected by the molecular size, the larger sized molecules diffusing less rapidly than smaller ones and liquids diffusing less rapidly than gases.

OSMOSIS

Consider now a membrane that does not allow the passage of all molecules, i.e. a semipermeable membrane, separating a solution from a solvent as shown in Fig. 7.8.

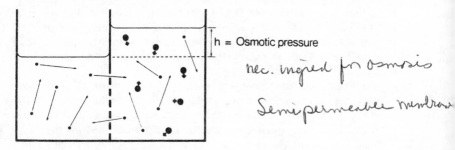

Figure 7.8 Principle of osmosis.

The small molecules of solvent can readily pass through the membrane, but in the compartment on the right there are some large solute molecules which cannot pass through. In this compartment the concentration of small solvent molecules is reduced by the presence of the solute. If Fick's law of diffusion is applied, what would happen to the small molecules? They diffuse from the area of higher concentration on the left across to the right side, giving a transfer of liquid. If the transfer is allowed to continue long enough, equilibrium is achieved and then 'h', the excess height of the compartment on the right, represents the osmotic pressure. It is the pressure which exactly counterbalances the effect of the dissolved molecules. An alternative way of achieving equilibrium is to have a closed compartment as shown in Fig. 7.9.

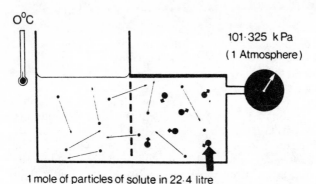

1 mole of particles of solute in 22·4 litre

Figure 7.9   101·325 kPa osmotic pressure is produced when 1 mol solute is dissolved in 22·4 litre of solution at 0 °C.

If the compartment on the right has a 22·4 litre capacity, and if it contains 1 mol of particles of solute at a temperature of 0 °C, then the osmotic pressure built up in this compartment is 101·325 kPa or one standard atmosphere. This is shown in the diagram by the gauge pressure, this being additional to the ambient pressure also present. Note that this is similar to the concept that 1 mol of a perfect gas in a volume of 22·4 litres exerts a pressure of one atmosphere (Chapter 4). Provided that 1 mol of solute is present in the compartment and that the membrane is not permeable to the solute molecules, the size of these molecules is immaterial.

In clinical practice there are many different types of molecules present in plasma, urine or other body fluids. In order to distinguish between the molarity of individual specific components and the sum total of molarities which give rise to the osmotic pressure the term 'osmolarity' is used. As an example consider the components of Ringer lactate solution.

Ringer lactate solution:

Sodium—131 mmol litre^{-1}
Potassium— 5 mmol litre^{-1}
Calcium— 2 mmol litre^{-1}
*Lactate— 29 mmol litre^{-1}
Chloride—111 mmol litre^{-1}

Total osmolarity—278 mosmol litre^{-1}

* Assuming the lactate is all in ionised form.

Each component is present in various concentrations of mmol litre^{-1}, but it is the sum total of the molarities which gives rise to the osmotic pressure. In other words, the osmolarity of the solution is the sum of the molarities of the solutes. In Ringer lactate solution it is 278 mosmol litre^{-1}.

Most body fluids such as plasma have an osmolarity of about 300 mosmol litre^{-1}. Over 99% of the osmolarity of plasma is due to electrolytes such as sodium, chloride and bicarbonate, the contribution of the plasma proteins being very small at around 1 mosmol litre^{-1}. Most fluids administered to the patient are adjusted to have an osmolarity of around 300 mosmol litre^{-1}. This means that they are isotonic with body fluids. If a patient is transfused with fluids of low osmolarity, i.e. hypotonic fluids, then the change in the osmotic pressure gradient across the cell membranes causes fluid to diffuse into the cells giving a rise of hydrostatic pressure in them.

Use is made of this fact in the red-cell fragility test for the detection of haemolytic anaemias. Red cells are added to various concentrations of saline at 20 °C. The lower the osmolarity of the solution, the greater the hydrostatic pressure which builds up in the red cells as fluid passes into them. Thus, at osmolarities below 200 mosmol litre^{-1}, the red cells burst and release haemoglobin into the solution. In many haemolytic anaemias the red cells are abnormal and more fragile and so burst at higher osmolarities than normal.

The main difference between plasma and interstitial fluid is the relative proportions of protein, which is negligible in the latter. As the membrane of blood capillaries is permeable to water and electrolytes but not to large protein molecules, the osmotic pressure between the blood and extracellular fluid depends on the protein molecules. To distinguish this pressure from total osmotic pressure it is sometimes called the oncotic pressure. Although albumin and globulin in the plasma have a very small osmolarity of around 1 mosmol litre^{-1} they give rise to an oncotic pressure in the capillaries of about 3·5 kPa (26 mmHg), the pressure being principally due to the albumin because of its higher molar concentration.

Figure 7.10 illustrates a capillary loop with an arterial pressure of 7 kPa (53 mmHg) and a venous pressure of near zero. The balance of pressures is closely related to the oncotic pressure from the plasma proteins. Thus, at the arterial end of the capillary, the hydrostatic pressure exceeds the oncotic pressure by 3·5 kPa and this results in the passage of fluid out of the capillary. At the venous end the pressure ratios are reversed and tissue fluid therefore passes into the capillary. These pressure gradients produce a continual fluid flow through the tissues to help carry oxygen and carbon dioxide and other agents to and from the cells. This is an important mechanism because diffusion alone does not enable solutes to move rapidly through a liquid. The example is simplified as the small hydrostatic and oncotic pressures in the tissue fluid have been disregarded.

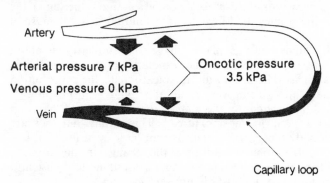

Figure 7.10   Pressure balance in a capillary loop.

A fall of plasma oncotic pressure, e.g. to below 2 kPa (15 mmHg), lowers the gradient at the venous end of the capillary and consequently tissue fluid can then accumulate as oedema. Measurement of oncotic pressure may therefore help with the management of patients with pulmonary or tissue oedema. The instrument used for such measurements is an oncometer and is based on the principle shown in Fig. 7.9. A semipermeable membrane separates the plasma sample from a saline reference solution, and the change due to oncotic pressure is measured by a transducer.

OSMOLARITY AND RENAL FUNCTION

In the kidney, filtrate passing from the glomerulus is concentrated by the cells in the renal tubules, as illustrated diagrammatically in Fig. 7.11. The plasma osmolarity is about 300 mosmol litre^{-1} and so the filtrate reaching the tubules is also close to this osmolarity. As the

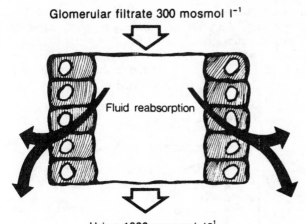

Glomerular filtrate 300 mosmol l^{-1}

Fluid reabsorption

Urine 1000 mosmol l^{-1}

Figure 7.11 Osmolarity in a renal tubule.

contribution of plasma proteins to the total osmolarity of plasma is very slight compared with that of the electrolytes, their absence from the filtrate is not an important factor. In the tubules active reabsorption, especially of water, then takes place and the urine finally produced may have an osmolarity of about 1000 mosmol litre^{-1}, three to four times greater than the plasma osmolarity.

The relative osmolarity of the urine to that of the plasma gives a good index of the efficiency of renal function. If the urine osmolarity has fallen to that of the plasma, this suggests either that the patient is having a diuresis or that there is some impairment of renal function. In other words a urine–plasma osmolarity ratio of just over 1 with a low urine output suggests an impairment of renal tubular function.

In reports of urine osmolarity an alternative expression, osmolality, is commonly used. Osmolality refers to the number of osmoles per kilogram of water or of other solvent. In scientific research, osmolality is the preferred term because it avoids inaccuracies from the effect of temperature on the volume of the solution. The analogous terms, molarity and molality have a similar meaning of moles per litre of solution and moles per kilogram of water (or other solvent) respectively.

MEASUREMENT OF OSMOLALITY

To measure the osmolality of urine or of plasma, an osmometer is used. Its principle is based on the depression of the freezing point of a solution, a phenomenon related to osmosis. This phenomenon is also the basis of the use of salt to melt ice on paths and roads. The depression of freezing point of a solution is directly proportional to its osmolality and 1 mol of a substance added to 1 kg of water depresses the freezing point by 1·86 °C. Similarly, the depression of the freezing point of the water in plasma and urine is related to the osmolality and this principle is used in the osmometer, illustrated in Fig. 7.12.

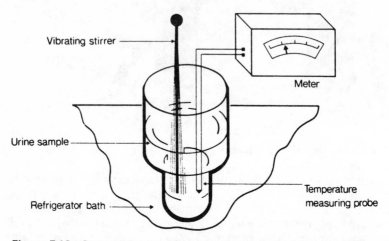

Figure 7.12   Osmometer based on the depression of the freezing point.

Two millilitres of urine are put in the sample tube in the refrigerator bath and a measuring probe senses the temperature to give a reading on the meter. A vibrating stirrer helps to mix the solution and to start even freezing. In practice, the urine cools to below the temperature of its freezing point before actually solidifying and then warms to a steady freezing point temperature as the solution changes state. This steady reading can then be noted. To avoid the need to convert freezing point depression to milliosmoles, modern instruments are calibrated to give a direct indication of osmolality when the meter recording the temperature finally achieves a steady reading.

Depression of the freezing point is not the only effect of a dissolved substance.

Figure 7.13 shows a pure solvent on the left and a solution with a solute present in the compartment on the right. Arrows above the liquid indicate that some solvent molecules vaporise to give rise to a vapour pressure above. On the right, the large solute molecules of the substance reduce the surface area available for the escape of the smaller solvent molecules. The solute therefore reduces the vapour pressure of the solvent. This effect is known as Raoult's law.

Raoult's law states that the depression or lowering of vapour pressure of a solvent is proportional to the molar concentration of the solute.

An osmometer using this principle of lowering of vapour pressure by a solute is also available. It has the advantage over osmometers which use the depression of freezing point in that it uses a smaller sample and can thus be used to measure the osmolality of sweat. The osmolality of sweat is proportional to its total electrolyte content which is altered, for example, in patients suffering from cystic fibrosis.

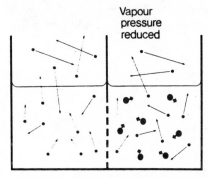

Vapour
pressure
reduced

Figure 7.13   Illustration of Raoult's law.

AZEOTROPES

Raoult's law applies to all solutions and the substance dissolved in solution need not be a solid or a gas but may be another liquid. A special example of one liquid dissolved in another is an azeotrope. *one will vaporize until air at a fixed propor* — An azeotrope is a mixture which vaporises in the same proportions as the volume concentrations of the components in solution. Ether and halothane form an azeotrope, provided that they are in the ratio of one part of ether to two parts halothane by volume. *Cannot purify one from another*

	Volume concentration	Mass concentration	Molar concentration
Ether	333 ml litre^{-1}	236 g litre^{-1}	3·19 mol litre^{-1}
Halothane	667 ml litre^{-1}	1240 g litre^{-1}	6·30 mol litre^{-1}

By application of the density and the molecular weights of ether and halothane respectively, the mass concentrations and molar concentrations may be calculated. As shown in the table, the ratio of the molar concentrations is 1 to 2 and so, by Raoult's law, the vapour pressure will also be in this same proportion. This means that the components of this azeotrope evaporate in the ratio of one part ether to two parts halothane and therefore the relative volume concentrations of the liquid mixture do not change. Azeotropes are found for other mixtures; thus alcohol in water forms an azeotrope at about 96% alcohol, making it impossible to prepare concentrations over 96% by fractional distillation.

Because the presence of another substance dissolved in a solvent lowers its vapour pressure and makes it less volatile, its boiling point is raised.

These phenomena—the depression of the freezing point, the lowering of the vapour pressure, and the raising of the boiling point—are all related to osmolarity and are sometimes known as the colligative properties of a solution.

Work, Energy and Power

WORK

This chapter concentrates on work, energy and power, with special reference to ventilation and cardiac output. Mechanical work is a form of energy, and other forms of energy include heat energy and chemical energy. Work is done or energy is expended whenever the point of application of a force moves in the direction of the force. The SI unit of energy is the joule.

> *One joule* of work is done when a force of one newton moves its point of application one metre in the direction of the force.

The force of gravity on an apple which has a mass of 102 grams is 1 newton, so if the apple is raised 1 metre vertically against the force of

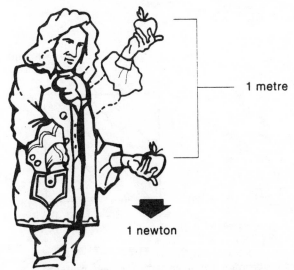

1 metre

1 newton

Figure 8.1 Illustration of the definition of the joule.

107

gravity, 1 joule of work is performed as illustrated in Fig. 8.1. The total energy must remain constant and, in this case, the mechanical energy expended in raising the apple is converted into potential energy represented by the additional height of the apple.

Forces in the body normally arise from muscular contraction; the shortening of the muscle multiplied by the mean force exerted is the mechanical work performed.

WORK OF A VENTILATOR DURING INSPIRATION

A constant-pressure generator type of ventilator (Fig. 8.2) gives a good example of the inter-relationship between the force and distance moved, and the pressure and volume changes. In this example, the force of gravity F on mass M acts over the area A to give a pressure P. During the descent of the bellows through a distance D, a volume of air V is delivered.

Work done $= F \times D$

but $\quad P = \dfrac{F}{A}\quad$ therefore $F = PA$

and $\quad V = D \times A\quad$ therefore $D = \dfrac{V}{A}$

Consequently the work done $= PA \times \dfrac{V}{A}$

$$\text{work done} = PV$$

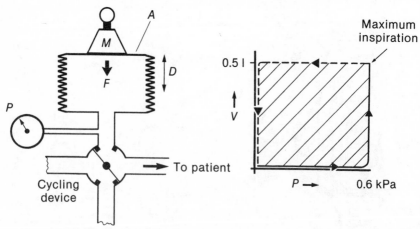

Figure 8.2   Work of inspiration for a perfect constant-pressure generator ventilator.

It can be seen that the work done is the product of the volume of gas moved and the pressure required.

Figure 8.2 also shows the volume–pressure graph of the inspiration, and the shaded rectangular area equals PV or the work done by the perfect constant-pressure generator ventilator in one inspiration. In this example the pressure from the ventilator increases rapidly to 0·6 kPa (6 cmH$_2$O) above atmospheric pressure and remains steady at this value until the lungs are inflated with 0·5 litre air. The work done by the ventilator can be calculated in SI units, provided that pressure is expressed in pascals and volume in cubic metres (1 m^3 = 10^3 litre).

$$\text{Work done} = (0 \cdot 6 \times 10^3) \text{ Pa} \times (0 \cdot 5 \times 10^{-3}) \text{ m}^3$$
$$= 0 \cdot 3 \text{ J}$$
$$= 300 \text{ mJ}$$

Thus, this inspiration requires 300 mJ of work.

During expiration the pressure at the ventilator falls rapidly to atmospheric and remains at zero while the air is expired. The dotted line on the graph shows the pressure and volume changes during expiration.

For simplification, a constant-pressure generator has been chosen for this example, but few ventilators act as perfect constant-pressure generators.

Most ventilators give a smoother build-up of pressure as indicated in Fig. 8.3, but the concept that the shaded area enclosed by the loop indicates the work of the ventilator for one inspiration is still valid. In this case, the calculation of the area is more complex and uses the process of integration (considered in Chapter 25).

Figure 8.3 Volume–pressure graph for inspiration and expiration with a typical ventilator.

ENERGY CHANGES

Energy cannot be lost but is converted from one form to another. During inspiration about half the mechanical energy used is stored as potential energy in the elastic tissues of the lung and the chest wall. This energy is subsequently used for the work of expiration. The remaining half of the mechanical energy of inspiration is used in overcoming airway resistance and in moving the air and tissues.

Figure 8.4 illustrates a model to help understand the energy changes in inspiration. A ventilator of a constant-pressure generator type is attached to the patient via a cycling device. The patient's lungs are represented by a syringe in which movement of the plunger is opposed by the stretching of an elastic band. In this model the plunger represents the diaphragm and the elastic band represents the elasticity of the lung tissues and chest wall. The resistance to the flow of air into the syringe represents the airway resistance and the pressure within the syringe is indicated by P_L. It is assumed that the end-inspiratory pressure in the syringe reaches the pressure P_M provided by the ventilator.

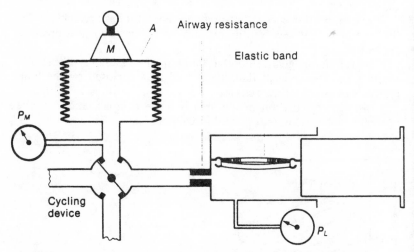

Figure 8.4 Simple syringe model to illustrate inspiration in controlled ventilation. Elastic band represents the elasticity of the lung and chest wall.

As previously explained, the total work done by the constant-pressure generator ventilator for one inspiration is represented by the rectangular area PV in the graph (Fig. 8.5). This energy can be divided into two parts; energy used in stretching the elastic band, and energy used in overcoming resistance to airflow into the syringe.

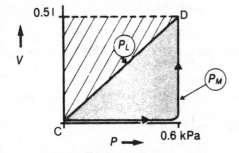

Figure 8.5 Graph of pressure change P_L in the model lung syringe during inflation with a volume of 0·5 litre. The ventilator pressure P_M is also indicated.

To assess the energy used in stretching the elastic band, consider how the pressure P_L changes within the syringe. This could represent the pressure changes within the alveoli. The pressure generated by the elastic band is atmospheric at the start of inspiration and increases linearly with volume as the air flows into the system to give a tracing C to D, the slope of which represents compliance. The average pressure is half the end-inspiratory pressure P, so the work done against the elastic band is $\frac{1}{2}PV$ and is equal to the area of the shaded section of the volume–pressure graph bounded by the line C to D.

The difference between this area and the total area PV is shown in the stippled section of the volume–pressure graph and represents the energy expended in moving the gas against the airway resistance. In the example at the start of inspiration, P_L is zero, as the syringe has been open to atmospheric pressure, and the ventilator pressure P_M is 0·6 kPa, so the initial pressure gradient across the airway resistance is 0·6 kPa. This gradient decreases to zero at the end of inspiration and so the average pressure difference is again $\frac{1}{2}P$ and the energy expended in moving the gas is $\frac{1}{2}PV$.

WORK OF EXPIRATION

The syringe model (Fig. 8.4) may also be used to illustrate expiration. The cycling device is open to the atmosphere and the pressure within the syringe then falls linearly to zero if plotted against the volume expired. The work expended is consequently $\frac{1}{2}PV$, and is used in overcoming the airway resistance in the model. In patients, the tissue resistance of the lungs and chest wall is present in addition to airway resistance, but is not illustrated in the model.

WORK OF INSPIRATION IN SPONTANEOUS BREATHING

In a quiet inspiration the contraction of the diaphragm muscle draws down the diaphragm by about 15 mm, causing a reduction of pressure within the chest and the inspiration of air. To represent this, the syringe model must be modified to that shown in Fig. 8.6. The syringe now has two plungers with fluid between them. The fluid represents the thin layer of serous fluid in the intrapleural space, while the right-hand plunger represents the diaphragm and the left-hand plunger represents the lung tissues moved.

The pressure in the intrapleural space is subatmospheric due to the elasticity of the band and decreases further as a result of the contraction of the diaphragm, represented by the force shown on the right-hand plunger. Through the fluid coupling of the intrapleural space this force is transferred to the lungs and inspiration occurs. A plot of volume against intrapleural pressure for a typical inspiration is shown in Fig. 8.6. As in the case of controlled ventilation, the stippled area on the graph gives the work expended in overcoming airway resistance and the triangular shaded area gives the stored elastic energy which is used during expiration.

If in this model 0·5 litre of air is inspired when the average pressure in the coupling fluid is 0·6 kPa, then 300 mJ of work is done in inspiration.

This simple syringe model does not show the chest expansion which also occurs, and which accounts for 10 to 40% of the inspired tidal volume. This expansion is caused by three factors: the diaphragm acting on the lower margin of the rib cage to raise it; the intercostal muscles, especially the external intercostals acting to raise the ribs; and the scalene muscles raising the upper ribs of the thoracic cage.

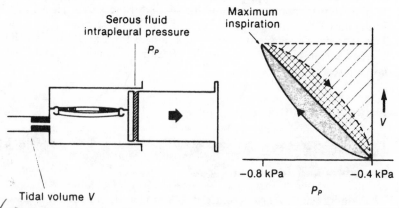

Figure 8.6   Model to illustrate spontaneous inspiration. The space between the two plungers represents the intrapleural space.

The initial energy for the contraction of the inspiratory muscles originates in the high-energy bonds of adenosine triphosphate (ATP), and most of this energy has in turn arisen from that released during the combination of oxygen with water and electrons in the respiratory chain of enzymes.

$$O_2 + 2H_2O + 4e = 4(OH)^- + Energy$$

In the respiratory muscles 10% of the chemical energy is turned into mechanical energy and 90% into heat; Fig. 8.7 summarises the energy changes for a typical inspiration and expiration.

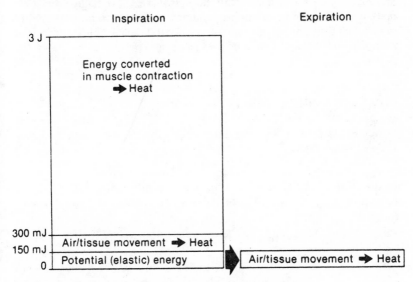

Figure 8.7   Energy changes in inspiration and expiration.

MEASURING THE WORK OF INSPIRATION

In controlled ventilation, analysis of a graph of airway pressure against volume change gives the work of inspiration, but the total work of inspiration is more difficult to measure in a spontaneously breathing patient. Although the volume inspired may be measured by a pneumotachograph (Chapter 3), the pressures exerted by the respiratory muscles in the chest wall cannot be monitored directly. Intrapleural pressure changes, however, indicate pressure changes acting on the lung, and the pressure in the lower oesophagus can be measured by a catheter with a balloon at its tip and used as an indication of these pressure changes.

Figure 8.6 illustrates the graph of volume against pressure which might be obtained in this way, but in this example the resistive and elastic work against the tissues of the chest wall are not measured. During expiration the graph traces the curve shown by the dotted line. The greater the expiratory resistance the greater the bowing of this line to the right; the cross-hatched area to the left of this dotted line represents the expiratory work done against airway and lung-tissue resistance while the balance of the stored energy in the triangular area is converted into heat in the tissues.

This technique has its limitations; not only does it exclude the work of movement of the chest wall but also the oesophageal pressure readings cannot give absolute accuracy. This is because they can indicate pleural pressure at only one site and the pleural pressure varies according to both the site of measurement and the posture of the patient.

POWER OF BREATHING

Power is the rate of working and is measured in watts, 1 watt being 1 joule per second.

$$W = J\,s^{-1}$$

In the earlier example the work of one inspiration was 300 mJ. If the respiration rate is 16 per minute, the power of breathing is calculated as follows:

$$\text{Mechanical power of breathing} = 300\,\text{mJ} \times \frac{16}{60}\,s^{-1}$$

$$= 80\,\text{mW}$$

EFFICIENCY OF RESPIRATORY MUSCLES

The respiratory muscles are only 10% efficient in producing mechanical energy, the rest of the energy being dissipated as heat. Thus, the actual energy requirements in the example above would be ten times greater than 80 mW—i.e. about 800 mW. As the normal metabolic rate is about 80 W under resting conditions (Chapter 9), in this example the energy requirements for breathing would be 1% of the total metabolism.

THE EFFECT OF THE TYPE OF FLOW

The power of breathing depends on the type of flow in the air passages, as pressure gradients are greater in the presence of turbulent flow than with laminar flow (Chapter 2). Consider, for example, the energy used as fluid flows through a tube. As the product of average pressure difference and volume moved indicates the energy, pressure difference across a tube multiplied by the fluid flow gives the power, if the pressure difference is small compared with atmospheric pressure.

$$\text{Energy } E = P \times V$$

Therefore

$$\text{Power } \dot{E} = P \times \dot{V}$$

In the case of laminar flow, pressure is proportional to flow:

$$P_L \propto \dot{V}_L$$

Therefore   Power in laminar flow $\dot{E}_L \propto \dot{V}_L{}^2$

In the case of turbulent flow, pressure is proportional to the square of the flow.

$$P_T \propto \dot{V}^2$$

Therefore   Power in turbulent flow $\dot{E}_T \propto \dot{V}_T{}^3$

It is seen that the power dissipated as fluid flows through a tube varies between the square and the cube of the flow rate.

The above calculations are simplified in that they do not allow for the kinetic energy in the flowing fluid, but in the flow rates normally present in the respiratory and circulatory systems the kinetic energy is very small.

EFFECT OF HYPERVENTILATION IN PATIENTS

In quiet ventilation, flow is predominantly laminar but becomes turbulent as ventilation increases, so that during hyperventilation the power of breathing increases rapidly at a rate approaching the cube of the gas flow. As the gas flow in the airway is more or less proportional to the tidal ventilation, the power of breathing increases rapidly with hyperventilation at a rate approaching the cube of the tidal ventilation, and a high oxygen consumption by the respiratory muscles occurs.

In some respiratory diseases the oxygen requirements of the respiratory muscles during hyperventilation become considerable and exceed the extra oxygen which is transported by the increased ventilation. Hence, pharmacological stimulation of ventilation in respiratory disease carries the risk of hypoxia in such patients.

LUNG COMPLIANCE AND AIRWAY RESISTANCE

Detailed consideration of these factors is beyond the scope of this book, but it should be noted that the power of respiration varies with the respiratory rate and that for a given patient there is an optimum respiratory rate depending on the compliance of the lungs and on airway resistance.

The time-constant for the lungs is the product of compliance and resistance (Chapter 5). Patients with a low time-constant tend to have a rapid respiratory rate whereas a high constant is associated with a slower rate.

WORK OF MYOCARDIAL CONTRACTION

The area of a graph of the changes in cardiac pressure and volume can be used to measure the work of myocardial contraction in the same way that the area of a pressure–volume loop in ventilation can be used to indicate the energy changes in breathing.

Figure 8.8 shows a tracing of the volume and pressure changes in the left ventricle during a typical cardiac cycle. The pressure and volume axes are reversed from those given when ventilation was discussed, as this is the more usual presentation when considering cardiac physiology. In this example, the left ventricular volume increased 60 ml during diastolic filling, then the pressure increased from 0 to 16 kPa (0–120 mmHg) during isovolumetric contraction. The systolic ejection

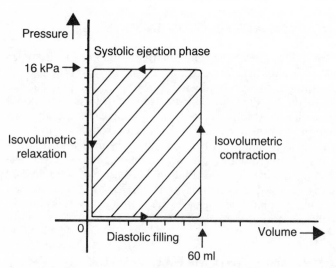

Figure 8.8   Graph of the volume and pressure in the left ventricle during a typical cardiac cycle.

phase and isovolumetric relaxation are indicated on Fig. 8.8, and the area of the loop represents the work done.

$$\text{Work done} = (16 \times 10^3)\,\text{Pa} \times (60 \times 10^{-6})\,\text{m}^3$$
$$= 960 \times 10^{-3}\,\text{J}$$
$$= 960\,\text{mJ}$$

Thus one myocardial contraction of the left ventricle requires just under one joule of work. If the heart rate is 60, then the power of the left ventricle of the heart is about 60 joules per minute, i.e. one watt.

To make such measurements of work, continuous ventricular pressure recording may be obtained from an intraventricular cardiac catheter, and volume measurements may be deduced from ultrasonic flow measurement (Chapter 13) or estimated from echocardiography (Chapter 13), cineangiography or other imaging techniques.

POWER OF THE HEART

The concept of power as a product of pressure difference and fluid flow may also be applied to the cardiac output. For example, if the mean arterial pressure is 12 kPa (90 mmHg), the pulmonary venous pressure 0 kPa and the left cardiac output 5 litre min^{-1}, then the power of the left heart ($\dot{E}$) can be calculated as follows:

$$\dot{E} = P \times \dot{V}$$
$$= (12 \times 10^3)\,\text{Pa} \times \left(\frac{5 \times 10^{-3}}{60}\right)\,\text{m}^3\,\text{s}^{-1}$$
$$= 1\,\text{W}$$

The power of the right heart can be similarly calculated as follows. If the mean pulmonary artery pressure is 2·4 kPa (18 mmHg) above the central venous pressure, then:

$$\text{Power of right heart} = (2\cdot4 \times 10^3)\,\text{Pa} \times \left(\frac{5 \times 10^{-3}}{60}\right)\,\text{m}^3\,\text{s}^{-1}$$
$$= 0\cdot2\,\text{W}$$

$$\text{Total power of heart} = 1\cdot2\,\text{W}$$

If the efficiency of the heart were 15% in this instance, then the total energy requirements of 8 W would be 10% of a typical basal metabolism of 80 W.

Energy is also generated by the heart to provide the kinetic energy of the flowing blood but this is a relatively small component of the total power.

To make the measurements for the above example the mean arterial pressure could be obtained from an arterial pressure trace and pulmonary pressures from cardiac catheterisation, while cardiac output could be measured by a dye dilution technique (Chapter 5) or by a technique based upon the Fick principle (Chapter 3).

EFFECT OF BLOOD PRESSURE AND CARDIAC OUTPUT ON THE WORK OF THE HEART

As power is the product of pressure and flow, the work of the heart is directly proportional to the mean blood pressure and to the cardiac output.

Hypertension results in additional energy requirements even if the cardiac output is unchanged, and hypotension lowers the energy demands on the heart provided that cardiac output is not simultaneously raised. A high cardiac output due to any cause (e.g. thyrotoxicosis, anaemia or exercise) also increases the work load of the heart. An increase in the energy requirements of the heart, if maintained, may lead to heart failure.

Temperature

THE CONCEPT OF TEMPERATURE AND HEAT

Heat is a form of energy that can be transferred from a hotter substance to a colder substance, the energy being in the form of the kinetic energy of the molecules of the substance. Temperature is the thermal state of a substance which determines whether it will give heat to another substance or receive heat from it, heat being transferred from the substance at the higher temperature to the substance at the lower temperature. As described in Chapter 4, the temperature of a gas is a measure of the speed at which the molecules are moving.

An analogy may be drawn between the relationship of heat to temperature and the relationship of the quantity of a solute to its concentration. Thus, in the same way that temperature rises as heat energy is added to a substance, so concentration rises as a solute is added to a solution. In each case, if the substance or solution is divided into two equal parts, each part will have the same temperature or concentration although the heat energy or quantity of solute in each case is halved.

TEMPERATURE SCALES

When heat energy is added to a substance, not only does its temperature rise, but changes in its physical properties also occur. For example, the substance may expand, and its electrical resistance may change. Use is made of such phenomena in temperature measurement and in the construction of temperature scales. Thus, mercury which expands and contracts as the temperature changes, was the substance used by Fahrenheit to construct the first temperature scale.

Normally, temperature scales have arbitrarily chosen fixed points, one example of which is the triple point of water, the temperature at which ice, water and water vapour are all in equilibrium.

In the SI system the unit of temperature is the kelvin, which is defined as the fraction $1/273 \cdot 16$ of the thermodynamic temperature of the triple point of water. The Celsius scale is also approved and the intervals on this scale are identical to those on the kelvin scale (K), but the relationship between the two is as follows:

$$\text{Temperature (K)} = \text{Temperature (°C)} + 273 \cdot 15$$

Thus on the Celsius scale, the triple point of water occurs at a temperature of 0·01 °C.

NON-ELECTRICAL TECHNIQUES OF TEMPERATURE MEASUREMENT

The mercury thermometer, which utilises the change in volume of mercury with temperature, is a reliable system, used both in laboratories and in hospital wards. An advantage in clinical use is that it can be made in maximum reading form, achieved by an angulated constriction at the lower part of the mercury column which splits the mercury column after it has reached its maximum reading (Fig. 9.1A). This prevents the mercury above from contracting into the bulb until the thermometer is shaken.

Maximum reading thermometers may also be made by including a small metal index above the surface of the mercury as shown in Fig. 9.1B. When the mercury column contracts, this index is left behind at

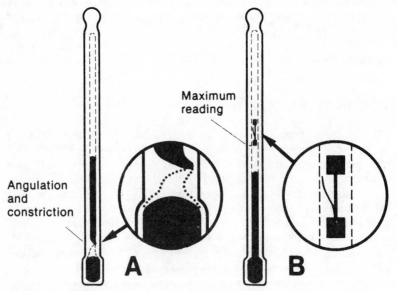

Figure 9.1   Maximum reading thermometers.

the level of maximum temperature. Mercury thermometers may also be used as thermostats by means of metal contacts inserted through the glass. A thermostat is a switch which operates at a preset temperature, and is often used to control electric power to a heater in order to maintain a constant temperature.

The mercury thermometer has two main disadvantages in clinical practice. Firstly, two to three minutes are necessary for complete thermal equilibrium between the mercury and its surroundings and, secondly, it may be difficult to introduce the thermometer into some orifices because of its rigidity and the risk of it breaking, with consequent injury to the patient.

Alcohol is sometimes used instead of mercury in thermometers. It is cheaper and is more suitable for use at very low temperatures, because mercury solidifies at $-39\,°C$. On the other hand, these thermometers are unsuitable for high temperatures because alcohol boils at $78·5\,°C$; another disadvantage is that the scale of the alcohol thermometer tends to be less linear than that of the mercury type.

Dial thermometers may use either a bimetallic strip or a Bourdon gauge.

Figure 9.2 shows a bimetallic strip thermometer, the sensing element of which consists of two dissimilar metals fixed together in a coil. As the

Figure 9.2   Bimetallic strip thermometer.

temperature rises, the metals expand by different amounts and the coil tightens to move the lever clockwise over the scale.

The Bourdon gauge type of thermometer is shown in Fig 9.3. This gauge is really a device for measuring pressure, and is attached to a sensing element containing a small tube of mercury or a volatile fluid. Variation in temperature causes a volume or pressure change in the sensing fluid and this is recorded on the Bourdon gauge which is calibrated in units of temperature.

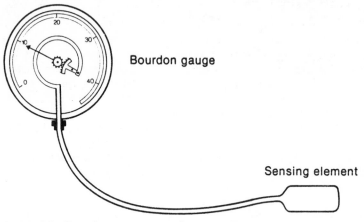

Figure 9.3   Bourdon gauge type thermometer.

ELECTRICAL TECHNIQUES OF TEMPERATURE MEASUREMENT

There are three principal electrical techniques for measuring temperature: the resistance thermometer, the thermistor, and the thermocouple.

The resistance thermometer is based on the fact that the electrical resistance of a metal increases linearly with temperature. A simple resistance thermometer could incorporate a platinum wire resistor, a battery as a source of electrical potential and an ammeter to measure current which could then be calibrated to indicate temperature. Unfortunately, a simple system of this type would not be very sensitive and it is better to incorporate the temperature-sensitive resistor into a Wheatstone bridge circuit containing an array of resistors. This circuit is dealt with in Chapter 14.

Another electrical thermometer system is based on the use of a thermistor. The usual type of thermistor is a little bead of metal oxide, the resistance of which—unlike the platinum resistance thermometer— falls exponentially as the temperature rises (Fig. 9.4). Special thermis-

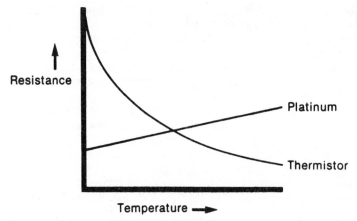

Figure 9.4 The graph contrasts the changes of resistance with temperature for a platinum resistance thermometer and a thermistor.

tors, however, are also available in which resistance rises with temperature. In the clinical temperature range the thermistor undergoes greater resistance change than the platinum resistance thermometer and, as with the latter, it is often used in a Wheatstone bridge circuit. The thermistor has the added advantage that the metal oxide bead can be made very small and that it can be manufactured more cheaply than the platinum resistance thermometer. A disadvantage is that the calibration is liable to change if the thermistor is subjected to severe changes of temperature, e.g. in heat sterilisation.

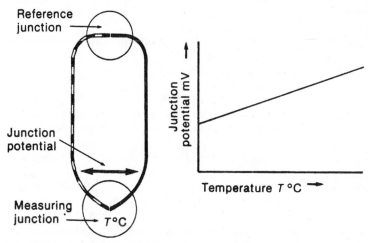

Figure 9.5   Seebeck effect. Principle of the thermocouple.

Another system of measuring temperature is based on the Seebeck effect, and the principle is shown in Fig. 9.5. At any junction of two dissimilar metals a small voltage is produced, the magnitude of which depends on the temperature of the junction. A junction of this type is known as a thermocouple, and metals such as copper and constantan may be used; constantan is an alloy of copper and nickel. A second junction is needed to form a complete electric circuit, but at this junction another temperature dependent voltage will develop. For the thermocouple to be used as a thermometer, one of the junctions, the reference junction, must be kept at a constant temperature while the measuring junction acts as a temperature probe; alternatively the equipment may provide electrical compensation for changes in reference junction temperature. An advantage of the thermocouple is that the measuring probe can be manufactured and used in the form of a needle.

CLINICAL APPLICATIONS

There is a choice of many sites for temperature measurement in patients—for example, the mercury clinical thermometer may be used in the mouth, axilla or rectum, but there is a limitation to its use owing to its rigidity. Rectal temperature measurement is of less value than measurements from other sites because, apart from aesthetic considerations, there is a risk in small babies of perforation of the rectum while in adults rectal temperature measurements may be inaccurate owing to the slowness with which rectal temperature equilibrates with body temperature.

The measuring probes of electrical thermometers may be introduced into other sites such as the nose, ear or oesophagus and they may also be applied to the skin. The most popular site is the lower oesophagus; in this case it is important that the recording tip of the thermometer is not too high, because the temperature of the oesophagus may then be reduced owing to the proximity of the lower trachea which is cooled by inspired air. The ear has also been used as a site to measure body temperature, but the obvious drawback to this site is the risk of perforation of the drum.

Nasal temperature fluctuates widely because of the passage of inspired and expired air and this fact has been utilised in the design of respiratory monitors, as illustrated on the left of Fig. 9.6. The pulses from fluctuation of temperature over the thermistor bead are electronically processed to give the respiration rate. The nose may be used as an alternative to the oesophagus to measure body temperature in an anaesthetised patient intubated with a cuffed endotracheal tube—as

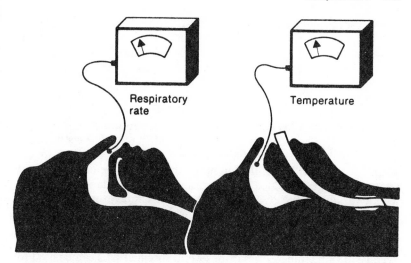

Figure 9.6   Use of nasal temperature to monitor respiratory rate or body temperature.

shown on the right of the figure—because there is then no respiratory air movement to cool the nasal cavity.

Skin temperature measurements are useful in anaesthesia because a fall of skin temperature in the periphery can be correlated with the degree of shock in an ill patient. A special system for measuring skin temperature is called thermography. In this technique the radiant heat from the patient's skin is imaged by means of an infra-red sensitive camera, to provide a thermograph of the relative warmth of the skin. Different colours in the thermograph indicate temperature differences and these, in turn, give an indication of the underlying vascularity. In this way, thermography has been used to detect tumours and vascular abnormalities.

BODY TEMPERATURE

Humans, like all mammals and birds, are homoeothermic and control their central body temperature within a narrow range despite a wide range of environmental temperature. In man, this range is $37 \pm 0.5\,°C$. A constant temperature is essential, as variation in temperature may bring about changes in enzyme reaction rates governing physiological processes and some metabolic processes could not occur if body temperature were allowed to cool to that of the environment.

Slight variation in body temperature does occur naturally in man.

There is a circadian or diurnal rhythm of around 0·4 °C, the temperature being lowest in the early morning and reaching a maximum in the early evening. There is also a slight variation in body temperature during the menstrual cycle in women, with the basal temperature rising after ovulation for the second half of the cycle.

As shown in Fig. 9.7, it is the central core of the body which is maintained at the constant temperature, while the surface layer or shell of the body is at a lower, more variable temperature which can range from 32 to 35 °C depending on physiological factors. The central core includes the brain, thoracic, and abdominal organs and also some of the deep tissues of the limbs, the amount varying with the environmental temperature as represented by the intermediate zone. The shell is a layer with a variable depth of around 2·5 cm. The temperature in the body core depends on the balance between heat production in the core and heat loss through the shell or surface layer.

In place of core temperature alone a formula combining this and an average of skin temperatures from several specified sites can be used to give the average patient temperature. Therefore:

Average patient temperature = 0·66 × Core temperature
+ 0·34 × Average skin temperature

In the formula it is assumed that two-thirds of the heat is in the core and one-third in the shell area. For example, if the core temperature is 37 °C and the average skin temperature is 34 °C, it can be calculated that the average patient temperature is 36 °C.

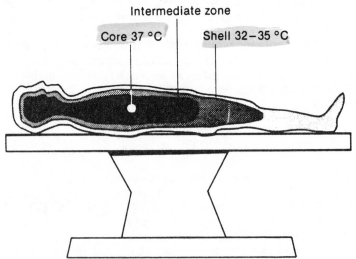

Intermediate zone

Core 37 °C   Shell 32–35 °C

Figure 9.7 Core and shell temperatures in the body. The intermediate zone represents variable core size.

HEAT PRODUCTION

Heat production is determined by the general metabolic rate of the person at any given time and is measured in watts. In an average man under resting conditions, heat production is of the order of 50 W m^{-2} body surface area or 80 W total, but increases of metabolic rate occur because of food consumption, exercise and other factors. In addition, an increase in metabolic rate is found when there is a rise of body temperature. There is no mechanism by which heat production can be reduced to compensate for overheating, but an increase of heat production can be provided by shivering.

Minor shivering is not apparent to the naked eye and may be detected only by an electromyogram. Major shivering is needed only when a large increase in heat production is required. Such shivering can more than double the metabolic heat production and a five-fold increase is possible in special circumstances. The onset of shivering does not normally occur until the body's mechanisms for the control of heat loss have already been activated.

HEAT LOSS

There are four routes of heat loss from the body as follows:

	Approximate % of heat loss	
Radiation	40	
Convection	30	
Evaporation	20	
Respiration	10	(8% evaporation of water; 2% heating of air)

It is seen that conduction *per se* is not important as a means of heat loss from the body. Conduction is the process whereby heat energy is transmitted through a substance by the transfer of the energy of motion of the molecules to adjacent molecules. Metals and some crystals are good conductors of heat but gases are poor conductors, and the air surrounding an individual protects him from heat loss from conduction unless excess air movement is present.

Heat loss from radiation is important and may account for up to 50% of normal heat loss from the body. A hot object emits radiation over a spectrum of wavelengths predominantly in the infra-red region. The radiation emitted carries energy from the hot object, thus causing it to cool down. Conversely, if this radiation is absorbed by another object, that object will become hotter. Thus radiation can transfer heat energy between two objects which are not in contact, the rate of transfer depending on the relative temperatures and the surface characteristics of the objects. The body acts as an almost perfect radiator. Radiant heat losses are increased if the body is surrounded by cold objects and reduced by radiation received from warm surroundings. Space blankets of shiny reflective metallised plastic foil are sometimes used to restrict radiant heat loss, but should not be used in theatre areas in view of the increased risks of burns and of electric shock (Chapter 16).

Convection is an important route of heat loss. In convection, the air layer adjacent to the surface of the body is warmed by conduction and as it is heated it expands and becomes less dense and so rises. The resultant convection current carries heat away from the subject.

Surface evaporative heat loss is due to the loss of latent heat of vaporisation of moisture on the skin's surface (Chapter 10). The loss of heat by this route is dependent on the water vapour pressure gradient from the skin to the air and to the total area of moist skin exposed to the atmosphere. The amount of heat lost by this route may be increased ten-fold by sweating, which increases the amount of moisture available for evaporation on the skin surface.

All these routes of heat loss depend on the total area of skin exposed to the environment. Consequently, if this area is reduced, heat loss is diminished. The rate of heat loss is also influenced by insulation, which constitutes a resistance to heat loss and arises from several sources. The body tissues lying between the core and the skin surface consist mainly of fat which forms a relatively avascular insulating layer. Clothing, too, acts not only by reducing the amount of skin surface area exposed but also by trapping air. So the trapped air, a poor conductor of heat, acts as an insulator.

Respiration accounts for only a small amount of the total heat loss, perhaps up to 10%. Eight per cent of this loss occurs in humidifying the inspired air and 2% in warming it. In humans, this route of heat loss is not an important mechanism for the physiological control of temperature, but changes in respiratory heat loss occur in anaesthesia. For example, inspiration of dry gases may contribute to the risk of hypothermia as explained in Chapter 10.

RESPONSES TO CHANGES IN ENVIRONMENTAL TEMPERATURE

Physiological control over temperature is mediated by temperature receptors in the skin and by a temperature receptor in the hypothalamus which is sensitive to the temperature of the blood perfusing it. Both types of sensor signal to a thermoregulatory centre in the hypothalamus and so initiate appropriate reflex changes. In the case of mild changes of environmental temperature, reflex changes of vascular tone occur and these alter skin temperature and so alter heat loss. In the case of major cooling of the body, reflexes elicit shivering while excessive heat causes reflex sweating and vasodilation. Signals from the thermoregulatory centre to the cerebral cortex lead to appropriate behavioural responses, so that the person concerned can alter the amount of clothing or his level of activity.

HYPOTHERMIA

The thermoregulatory centre is depressed by anaesthesia and many anaesthetics also have an effect on vasomotor tone giving peripheral vasodilation and thus increased heat loss. During surgery there is a greater area of the patient exposed to the atmosphere and there may be rapid air movement over him from modern theatre ventilation. In prolonged surgery, therefore, there is a risk of hypothermia and the patient's core temperature must be monitored and appropriate measures taken to maintain it—by raising theatre temperature, for example, by humidifying the inspired air (Chapter 10), and by warming the transfused blood.

Hypothermia may be induced intentionally in anaesthetised patients undergoing neurosurgery or cardiac surgery. The reduced body temperature decreases metabolism and thus the tissue demand for oxygen, so that blood flow to vital areas may be interrupted for a time without the risk of hypoxic damage. Thus at 30 °C, the time for safe circulatory occlusion is double that at 37 °C. In the simpler hypothermic techniques, core temperatures of below 30 °C are avoided because otherwise there is a risk of ventricular fibrillation, but lower temperatures are sometimes used in patients on extracorporeal circulation.

PYREXIA

In fever, the thermostat of the thermoregulatory centre may be regarded as being set at a higher level than normal, and this is usually due to a bacterial or endogenous pyrogen. Temperatures over 42 °C cause cerebral impairment and disturbance of the physiological control mechanisms of heat loss by sweating. At high body temperatures over 45 °C, the rise of temperature induces a rise in metabolism which leads to a vicious circle with even more heat being produced; the outcome is usually fatal. A problem at these high body temperatures is that heat production rises exponentially with temperature while the heat loss rises at a slower rate. Consequently, a temperature is reached above which heat production exceeds the maximum heat loss and the temperature then rises until death occurs.

A rare anaesthetic cause of such hyperpyrexia is the malignant hyperpyrexia which occurs in susceptible patients who have an abnormal response to halothane or suxamethonium. The condition is genetically determined and is associated with a dramatic rise in body temperature due to a massive increase in heat production carrying a serious risk of fatality.

THERMAL BURNS

There is a risk of burns and tissue damage if the temperature of the surface skin or tissues approaches 45 °C from an external heat source. The actual temperature reached depends on the rate at which heat arrives from the heat source, and the rate at which it is removed by the blood flow; consequently, burns are more likely in patients who are vasoconstricted or shocked. Examples of external heat sources are lamps, electric blankets and heated water-filled mattresses. Electrical burns are considered in Chapter 14 and the danger from fires in Chapter 23.

Heat Capacity and Latent Heat

The relationship between heat and temperature is considered in Chapter 9. When heat is applied to a substance its temperature rises and, conversely, when heat is removed from a substance, its temperature falls. The quantity of heat necessary to produce a given change in temperature depends on the specific heat capacity of a substance.

> *Specific heat capacity* is defined as the amount of heat required to raise the temperature of 1 kilogram of a substance by 1 kelvin (SI unit of specific heat capacity $J\,kg^{-1}\,K^{-1}$).

As 1 K is commensurate with 1 °C, the specific heat capacity is often quoted using the latter unit. The word 'specific' in a definition indicates that the quantity is expressed in terms of unit mass. Specific heat capacity is often abbreviated to specific heat and must not be confused with the term 'heat capacity'.

> *Heat capacity* is defined as the amount of heat required to raise the temperature of a given object by 1 kelvin (SI unit of heat capacity $J\,K^{-1}$).

It is seen that heat capacity applies to an object or substance as a whole; consequently, it can be calculated from the product of the specific heat capacity and the mass. An example of heat capacity is given in Chapter 11 when vaporisers are considered. If an object such as a vaporiser has several component parts of different materials, its heat capacity can be calculated by adding the capacities of the components. In the case of a person, the tissue specific heat capacities can be approximated to a mean value of $3.5\,kJ\,kg^{-1}\,°C^{-1}$ and the product of this and the patient's weight indicates his total heat capacity. For a 70 kg patient it would be $245\,kJ\,°C^{-1}$.

Knowledge of the heat capacity facilitates calculation of the effect of changes in heat production. Consider, for example, a patient after anaesthesia whose body temperature has fallen to 36 °C. By shivering he may increase his heat production four-fold to 320 W from a basal level of 80 W. This is an extra 240 W or 240 × 60 joules per minute, i.e. 14·4 kJ min^{-1}. As 245 kJ are needed to raise his temperature one degree, this patient will need to shiver for 17 minutes to produce the heat required to do this.

The concept of body heat content has been used to estimate the overall heat gain or heat loss of a person or an object. The body heat content is the product of mean temperature and heat capacity. In the case of a person there are temperature gradients between the core and the shell; consequently, both core temperatures and surface skin temperatures must be measured and a mean body temperature calculated allowing for the estimated sizes of core and shell tissues, as explained in Chapter 9.

The term 'thermal capacity' is also sometimes encountered, but is best avoided because it may mean either specific heat or heat capacity.

As shown in Fig. 10.1, 4·18 kJ are required to raise the temperature of 1 kg of water by 1 °C. The specific heat of water is therefore 4·18 kJ kg^{-1} °C^{-1}. In parenteral nutrition studies, however, calories and kilocalories may be used, 1 calorie being equal to 4·18 joules; so the specific heat of water is 1 kcal kg^{-1} °C^{-1}. It should be noted that in dietetics the kilocalorie is often written as Calorie with a capital C.

The specific heat of most other substances is less than that of water. The high specific heat of water is used in the design of vaporisers to provide a reservoir of heat and so maintain a steadier temperature. Thus, on older anaesthetic machines, a water reservoir was sometimes used around the Boyle's bottle vaporiser.

The high specific heat of cold infusion liquids and blood used in transfusions increases the cooling effect in the patient. The specific heat capacity of blood is 3·6 kJ kg^{-1} °C^{-1}.

Suppose 2 kg of blood, approximately 2 litre, are transfused at 5 °C and warmed up to 35 °C in the patient.

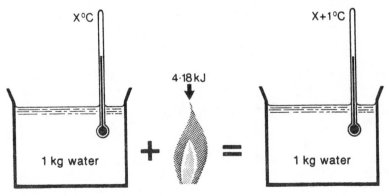

Figure 10.1   Specific heat of water.

The heat required $= 2 \text{ kg} \times 3\cdot6 \text{ kJ kg}^{-1}\,{}^{\circ}\text{C}^{-1} \times (35 - 5)\,{}^{\circ}\text{C}$
$$= (2 \times 3\cdot6 \times 30) \text{ kJ}$$
$$= 216 \text{ kJ}$$

The total heat capacity of a 70 kg patient is 245 kJ $^{\circ}\text{C}^{-1}$, as mentioned above, so the patient's mean temperature must fall by up to 1 °C when 2 litre of unwarmed blood are transfused. Blood warmers are therefore advised when transfusions of this size are required.

Gases have low specific heats and this may be important in anaesthesia. Air, for instance, has a specific heat of $1\cdot01 \text{ kJ kg}^{-1}\,{}^{\circ}\text{C}^{-1}$ at constant pressure. The quantity of a gas, however, is seldom measured in kilograms but usually expressed in terms of a volume. It is therefore more convenient to refer to the specific heat of a gas as the amount of heat required to raise the temperature of a unit volume of it by one temperature unit, i.e. the heat required to raise 1 litre of gas through 1 K (1 °C). The low densities of gases mean that the specific heats expressed in this way are very low, e.g. the specific heat of air is only $1\cdot2 \text{ J litre}^{-1}\,{}^{\circ}\text{C}^{-1}$, about a thousandth of the numerical value when expressed in terms of mass. The practical implication is that an extremely small quantity of heat is required or lost when the temperature of a small volume of a gas is altered. When a cigarette is smoked, for example, the burning tobacco is at a temperature of several hundred degrees Celsius. As air is drawn over this burning tobacco, it is heated to about this temperature. However, the temperature of this air falls dramatically by the time it reaches the mouth, even though the quantity of heat lost to the environment is small.

Another example, more appropriate to anaesthesia, concerns the temperature of the supply of anaesthetic gases to a patient. The gases at source may be quite cold if supplied direct from a vaporiser, or warm if supplied via a heated humidifier. As they pass along the anaesthetic tubing, only minimal quantities of heat are transferred, but this is sufficient to ensure that the gas is closer to ambient temperature by the time it arrives at the patient.

Normally the heat required to warm inspired air is quite small because of the low specific heat of the air. Consider a patient breathing at 7 litre min^{-1} with an endotracheal tube in the upper trachea. What is his heat loss from this source? The temperature in the upper trachea is about 34 °C and so, if any heat exchange effect within the endotracheal tube is ignored, the temperature rise of the air could be 14 °C, i.e. from a room temperature of 20 °C to 34 °C.

Heat loss from warming inspired air $=$ Flow $\times$ Specific heat capacity $\times$ Temperature rise
$$= 7 \text{ litre min}^{-1} \times 1\cdot2 \text{ J litre}^{-1}\,{}^{\circ}\text{C}^{-1} \times 14\,{}^{\circ}\text{C}$$
$$= 118 \text{ J min}^{-1}$$
$$= 1\cdot96 \text{ W } (1 \text{ W} = 1 \text{ J s}^{-1})$$

As shown in the calculation, the heat loss is only about 2 W, an insignificant amount compared with the patient's heat loss in basal conditions of about 80 W. Consequently, heat loss from warming inspired air is not normally an important factor but, if dry gases are being inhaled, extra heat is needed to supply the latent heat to humidify the air.

LATENT HEAT

When a substance changes from a liquid to a vapour or from a solid to a liquid, heat must be supplied even though this change of state takes place at a constant temperature. The heat which is required is known as the latent heat of vaporisation when the change is from a liquid to a vapour, or the latent heat of fusion when a solid changes to a liquid. Another category of latent heat is associated with a solid dissolving in or crystallising out from a liquid, and is known as the latent heat of crystallisation.

Of these forms, the latent heat of vaporisation is of most interest in anaesthesia.

✓ *Specific latent heat* is defined as the heat required to convert 1 kilogram of a substance from one phase to another at a given temperature (SI unit of specific latent heat $J\,kg^{-1}$).

Thus, water at 100 °C may be converted to steam at 100 °C by supplying the appropriate quantity of latent heat, i.e. 2·26 MJ kg^{-1} water (Fig. 10.2). Note that the temperature at which the change of state occurs must be specified, because the value of latent heat depends on this.

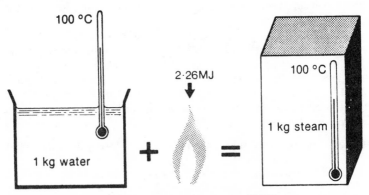

Figure 10.2   Specific latent heat of water at 100 °C.

At body temperature it is found that 2·42 MJ are required to turn 1 kg water into 1 kg vapour (Fig. 10.3). At room temperature the heat required is even greater, at 20 °C 2·43 MJ are needed, i.e. the lower the temperature the more latent heat is needed to vaporise a substance.

Figure 10.4 shows a graph of the change in latent heat with temperature for water. If this graph were extended to the right, the latent heat would continue to fall as temperature rose, and would ultimately reach zero. This is illustrated in the graph of the latent heat of vaporisation of nitrous oxide (Fig. 10.5).

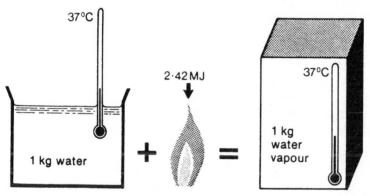

Figure 10.3 Specific latent heat of water at 37 °C.

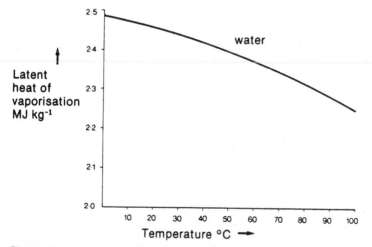

Figure 10.4 Graph of the specific latent heat of water against temperature.

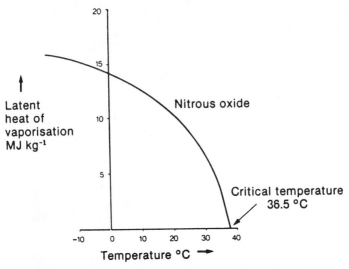

Figure 10.5 Graph of the specific latent heat of nitrous oxide against temperature.

The temperature at which the latent heat of vaporisation of nitrous oxide becomes zero corresponds with its critical temperature, 36·5 °C. At this temperature the substance changes spontaneously from liquid to vapour without the supply of any external energy. Above this critical temperature the substance cannot exist as a liquid. Critical temperature is discussed in Chapter 4.

LATENT HEAT IN ANAESTHESIA

A clinical example of latent heat of vaporisation is seen in the use of ethyl chloride for local anaesthesia. Ethyl chloride is stored as a liquid under pressure in glass ampoules with a tap so that a fine jet of the substance may be directed onto the skin. Vaporisation of the ethyl chloride causes pronounced cooling of the skin thus impairing conduction in the sensory nerves and providing enough analgesia for minor surgical procedures such as the opening of a skin abscess or whitlow.

All volatile anaesthetics lose latent heat when they are vaporised and this factor is important in the design of vaporisers. As gas passes over liquid anaesthetic in a vaporiser, the liquid vaporises taking latent heat of vaporisation from the remaining fluid and from the surrounding vaporiser walls (Fig. 10.6). Thus, the temperature of the remaining

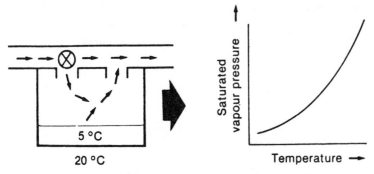

Figure 10.6 Effect of latent heat in a vaporiser.

anaesthetic agent and of the vaporiser walls falls. But a fall in temperature of the anaesthetic in the vaporiser renders it less volatile, lowering its saturated vapour pressure and so reducing the amount of anaesthetic vaporised.

Most modern vaporisers now have systems for controlling the concentrations of vapour based on thermostatic devices which are considered in Chapter 11.

Another effect of latent heat of vaporisation can be seen if a nitrous oxide cylinder is allowed to empty rapidly (Fig. 10.7). Nitrous oxide is

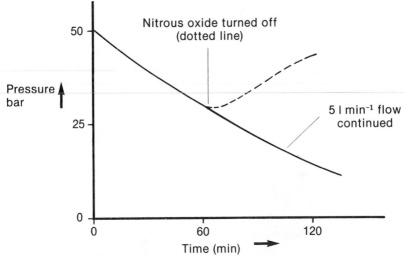

Figure 10.7 Changes in the pressure from a 900 litre nitrous oxide cylinder emptied at 5 litre min⁻¹. Pressure is regained if flow is turned off as indicated by the dotted line.

stored in cylinders in liquid form and, as it is used, it is converted to gas, the latent heat required being taken from the remaining fluid and from the cylinder walls. Thus the temperature of the cylinder falls and water vapour from the air may condense or freeze on the outside of it. Because of the cooling, the vapour pressure falls rapidly inside such a cylinder and a pressure gauge, if fitted, gives a low reading which recovers towards the previous level after the cylinder is turned off, as shown by the dotted line in Fig. 10.7.

Carbon dioxide and cyclopropane are also stored in liquid form, but in normal anaesthetic practice the rate of use is so slow that cooling of the cylinder is not noticed.

Oxygen is another example of a gas which may be stored in liquid form in a hospital, as this is more economical. Because the critical temperature of oxygen is $-119\,°C$, it cannot exist as a liquid above this temperature.

In practice, the liquid oxygen in the storage vessel is at a temperature of around $-160\,°C$, and must be kept in a special storage vessel resembling a gigantic vacuum flask to maintain it at this low temperature (Fig. 10.8). Normally the pressure inside the vessel is set to about 7 bar, this being the vapour pressure of the oxygen at $-160\,°C$.

When oxygen is taken from the top of the storage vessel, it is very cold and so it has to pass through a superheater coil, shown on the right of the diagram. Then it passes through a pressure regulator so that the pipeline pressure is kept at about 4·1 bar.

If oxygen flows at a fast rate, then the temperature of the liquid oxygen falls because of the removal of latent heat and its vapour

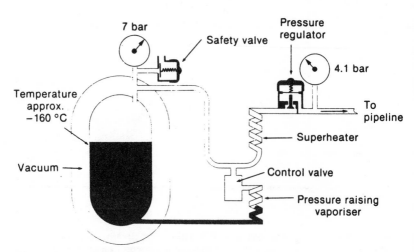

Figure 10.8   Liquid oxygen supply system.

pressure falls. To solve this problem, a supplementary source of heat is needed and this is provided by a pressure-raising vaporiser. A control valve senses the storage vessel pressure and controls the flow of liquid oxygen to the pressure-raising vaporiser—the lower the pressure, the higher the flow of liquid oxygen. In the vaporiser the oxygen is warmed and vaporised to the pipeline pressure.

No refrigeration unit is needed for the storage vessel because the liquid oxygen remains cold due to the efficiency of the vacuum container and the loss of latent heat as oxygen vaporises. Nevertheless, if no oxygen is used the temperature of the storage vessel gradually rises until after about a week the oxygen pressure causes excess oxygen to blow off through the safety valve. The evaporation of this liquid oxygen then reduces the temperature and pressure.

LATENT HEAT AND HEAT LOSS FROM THE PATIENT

The heat loss from warming inspired air has already been considered but not the heat loss for its humidification. By the time gases reach the trachea, their humidity is about 34 mg water vapour per litre of gas, this water coming from vaporisation from the mucosa of the respiratory tract. Consider now a person breathing completely dry gases at 7 litre min^{-1}.

Humidity in the upper trachea	$= 34$ mg litre^{-1}
Total water required to humidify inspired gas	$= 7$ litre min$^{-1} \times 34$ mg litre^{-1}
	$= 0.238$ g min^{-1}
Total latent heat needed	$=$ Specific latent heat of vaporisation at 37 °C × Total water
	$= 2.42$ MJ kg$^{-1} \times 0.000238$ kg^{-1} min^{-1}
	$= 576$ J min^{-1}
	$= 9.6$ W
Specific heat to warm inspired air (calculated previously)	$= 2.0$ W
Total heat loss from respiration	$= 11.6$ W

As shown in the calculations, the heat lost in humidifying the air is much greater than that lost in warming dry air, and the total heat loss from respiration of almost 12 W represents 15% of the total basal heat loss of about 80 W.

This is a higher figure than that given in Chapter 9, where it is stated that about 10% of the person's heat production is normally spent in warming and humidifying the inspired air. In anaesthesia, however, the inspired gases are usually dry and consequently greater heat is lost by this route than under normal conditions. This heat loss may contribute to the general problem of hypothermia during anaesthesia—especially in young children. It can be avoided by humidifying inspired gases and is reduced when an anaesthetic circle system with soda lime is used.

The older closed anaesthetic system with the Waters canister also allows efficient conservation of a patient's heat and moisture, and has even been used to induce controlled hyperthermia in one special anaesthetic technique. In this technique the patient is anaesthetised with the Waters canister system and other sources of heat loss are blocked by placing the person in a wax bath. By this means the patient's temperature may be raised as part of the treatment of cancer.

CHAPTER 11

Vaporisers

A vaporiser is a device for adding clinically useful concentrations of anaesthetic vapour to a stream of carrier gas. The saturated vapour pressure of volatile anaesthetic agents at room temperature is many times greater than that required to produce anaesthesia, and so a vaporiser mixes gas passing through a vaporising chamber with gas containing no vapour to produce a final mixture with the appropriate concentration. This is achieved by splitting the flow of gas to the vaporiser into two streams as shown in Fig. 11.1. One stream passes through the chamber containing the anaesthetic agent and the other bypasses it. These streams then reunite and pass to the patient.

Gas can be made to flow through a vaporiser in one of two ways. Firstly, a positive pressure can be developed upstream of the vaporiser, e.g. by gas from the flowmeter, so that gas is pushed through. This is known as a plenum vaporiser. Alternatively, as shown in Fig. 11.2, a negative pressure may be developed in the gas stream distal to the vaporiser, thus drawing gas through. This is known as a draw-over vaporiser and the negative pressure may be generated either by the patient's respiratory effort or by mechanical means. The plenum type of vaporiser is considered first.

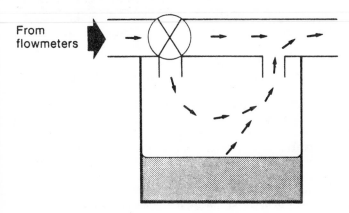

From flowmeters

Figure 11.1 Principle of the plenum vaporiser.

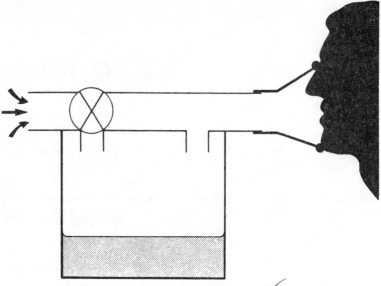

Figure 11.2　Principle of the draw-over vaporiser.

THE BOYLE'S BOTTLE

One of the earliest and simplest types of plenum vaporiser is the Boyle's bottle, illustrated in Fig. 11.3. The liquid anaesthetic is contained in a glass bottle and the proportion of gas flowing through the chamber relative to that flowing through the bypass is controlled by a rotary valve. In the illustration the control lever for this valve is shown in the full-on position, and in this case all the gas flows directly over the surface of the anaesthetic agent. If the control lever is rotated downwards, part of the gas flow is diverted through the bypass channel shown and, in this way, the splitting ratio, i.e. the fraction of gas passing through the vaporising chamber, may be varied.

In this vaporiser the degree of saturation of the gas leaving the chamber is very dependent on the flow rate, the final concentration being greater at low flow rates. In order to overcome this problem the vaporiser is fitted also with a plunger and cowl which can direct the gas stream closer to the surface of the liquid or enable it to be bubbled through the liquid agent. The vaporiser thus has the inconvenience of requiring a second control to operate it, but this plunger is needed during induction to achieve an adequate concentration of an anaesthetic agent such as ether. The Boyle's bottle is sometimes used for

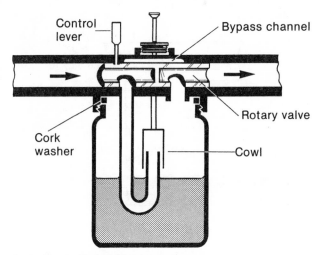

Figure 11.3   The Boyle's bottle.

trilene or other anaesthetics. In this case, high concentrations can be dangerous and so the plunger is raised to its fullest extent.

This simple vaporiser has another problem, discussed in greater detail in Chapter 10. As the anaesthetic agent vaporises, it draws latent heat of vaporisation from the remaining liquid anaesthetic and from the walls of the vaporiser. The temperature and hence the saturated vapour pressure of the agent therefore fall. The concentration of anaesthetic from the vaporiser then decreases unless readjustment of the settings is made. Because the output of the Boyle's bottle is so variable it cannot be calibrated.

To a large extent, developments in vaporiser design have solved the problems of flow and temperature dependence and more accurate models are now usually preferred to the Boyle's bottle. These vaporisers are normally specifically designed and calibrated for use with one anaesthetic agent only.

FLOW DEPENDENCE

The problem of flow dependence may be overcome if it can be arranged that all the gas emerging from the vaporising chamber is fully saturated at all flow rates so that the final percentage of anaesthetic vapour delivered from the vaporiser is controlled by adjustment of the splitting ratio and is independent of the flow.

Figure 11.4 illustrates a halothane vaporiser at a room temperature

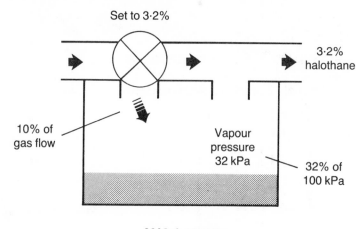

Figure 11.4   Halothane vaporiser at 20 °C illustrating principle of action.

of 20 °C. The saturated vapour pressure of halothane at this tempera-ture is 32 kPa. If the ambient atmospheric pressure is 100 kPa, i.e. 1 bar, the concentration of vapour in the vaporising chamber is 32%, an example of Dalton's law of partial pressures. If the control valve is set to allow 10% of the incoming gas to flow through the chamber and 90% through the bypass channel, the resulting concentration of halothane is 3·2%. The concentration remains the same provided that the tempera-ture is constant and the surface area of liquid halothane within the chamber is large enough to maintain the saturated vapour pressure.

The saturated vapour pressure of an anaesthetic such as halothane is greatly in excess of that required to maintain anaesthesia. If a fault occurs, e.g. in the control valve, then an unintentionally high propor-tion of gas may pass through the vaporising chamber and there is a risk of overdose and fatality. It is, therefore, essential for safety that the control valve and associated parts work reliably.

To achieve full saturation a large surface area must be available for vaporisation within the vaporising chamber. One method of achieving this is to place in the vaporising chamber metal or fabric wicks, one end of which dips into the anaesthetic liquid while the other end projects up into the chamber (Fig. 11.5A). Surface tension between the liquid anaesthetic and the capillary channels in the wicks draws up the anaesthetic and provides a large surface area of anaesthetic which assists efficient vaporisation. This method is used in most vaporisers, e.g. in the Abingdon, Dräger and Fluotec vaporisers.

An alternative method, shown in Fig. 11.5B, is to bubble the gas

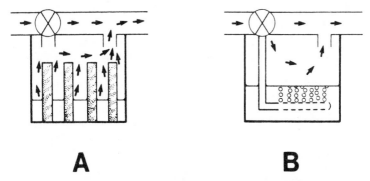

Figure 11.5   Techniques to achieve full saturation of the gas in the vaporiser chamber. (A) Use of wicks. (B) Use of small bubbles from sintered glass or metal.

through the liquid anaesthetic by means of a sintered disc of glass or metal. This produces large numbers of minute bubbles which have a large total surface area. The copper kettle vaporiser used in North America employs this system.

Despite the use of a technique with full vapour saturation in the vaporising chamber, circumstances can arise which affect a vaporiser's output. The splitting ratio produced in a particular vaporiser depends on the relative resistances to flow of the gas paths through the vaporising chamber and through the bypass channel. These resistances depend on whether the flow is laminar, turbulent or a mixture of the two, and on the physical properties of the gases (e.g. viscosity and density). Flow dependence is usually worst at low flows, e.g. below one litre per minute, and the manufacturer's chart must be consulted if low flows are used. Care should be taken to ensure that the conditions under which the vaporiser is calibrated represent the clinical conditions under which it is used. A vaporiser which is calibrated with oxygen flowing through it will generally give a slightly different output with nitrous oxide–oxygen mixtures.

The copper kettle series of vaporisers overcomes these problems by using a flowmeter to control the flow of gas through the vaporising chamber.

TEMPERATURE CONTROL

The glass in the walls of the Boyle's bottle and similar simple vaporisers is a poor conductor of heat and prevents effective conduction of heat from the surroundings. More advanced types of vaporisers are made of metal which has a good thermal conductivity and which allows heat to be transferred from the surroundings to the vaporising chamber. In addition, the metal itself acts as a reservoir of heat to delay and reduce temperature fluctuations as the heat capacity of the metal will be added to that of the anaesthetic in the vaporiser (Fig. 11.6). For example, consider a halothane vaporiser of 5 kg weight made of copper and containing 200 ml of halothane.

Heat capacity of halothane = Volume × Density × Specific heat capacity
$$= 200 \text{ ml} \times 1\cdot87 \text{ g ml}^{-1} \times 0\cdot8 \text{ J g}^{-1} \text{ K}^{-1}$$
$$= 299 \text{ J K}^{-1}$$

Heat capacity of copper $\quad = 5000 \text{ g} \times 0\cdot39 \text{ J g}^{-1} \text{ K}^{-1}$
$$= 1950 \text{ J K}^{-1}$$

Total heat capacity $\quad = 299 + 1950 \text{ J K}^{-1}$
$$= 2249 \text{ J K}^{-1}$$

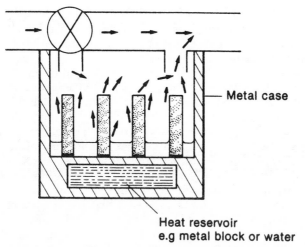

Figure 11.6 Partial control of the temperature in the vaporiser by the use of a metal casing and a heat reservoir.

Although the specific heat of copper is not as high as that of halothane, copper is very dense and the large mass present in a typical vaporiser gives an important contribution to the heat capacity.

Water has an even higher specific heat capacity and it can also be used as a heat reservoir. The concept of a water reservoir is not new, as traditionally a container of water has been used to surround the Boyle's bottle. If the Boyle's bottle contains 200 ml ether and the water reservoir is 200 ml, the calculations are as follows.

Heat capacity of ether = Volume × Density × Specific heat capacity
$$= 200 \text{ ml} \times 0.71 \text{ g ml}^{-1} \times 2.2 \text{ J g}^{-1} \text{ K}^{-1}$$
$$= 312 \text{ J K}^{-1}$$
Heat capacity of water $= 200 \text{ ml} \times 1 \text{ g ml}^{-1} \times 4.18 \text{ J g}^{-1} \text{ K}^{-1}$
$$= 836 \text{ J K}^{-1}$$
Total heat capacity $= 312 + 836 \text{ J K}^{-1}$
(excluding the glass
vaporiser bottle)
$$= 1148 \text{ J K}^{-1}$$

In practice, the thermal conductivity of glass is not as high as metal and this limits the temperature stabilising effect of the water bath.

Although the provision of heat reservoirs reduces the rate at which temperature changes occur, these changes are not eliminated.

TEMPERATURE COMPENSATION

A graph of the relationship of temperature and saturated vapour pressure for halothane is illustrated in Fig. 11.7. As mentioned above, at 20 °C the saturated vapour pressure is 32 kPa but at lower temperatures this value is smaller. Thus the output from the vaporiser falls unless the splitting ratio of the gas is altered so that more gas flows

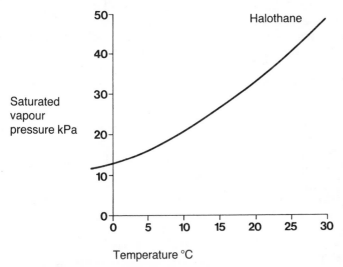

Figure 11.7 Graph illustrating the effect of temperature upon the saturated vapour pressure of halothane.

through the vaporising chamber. This adjustment, necessary to give a steady concentration, may be carried out manually or automatically.

In earlier models of the Dräger Vapor halothane vaporiser the process was done manually. A thermometer was provided to monitor the temperature in the vaporising chamber and a scale on the control knob allowed the anaesthetist to read accurately the percentage of halothane at a particular vaporiser temperature. This method gave accurate concentrations.

The copper kettle vaporiser also has a system of manual adjustment, but here the two gas flows, through the bypass and through the vaporising chamber, are entirely separate and are adjusted by separate flowmeters. The temperature in the vaporiser is noted and the flow rates through the two flowmeters are adjusted according to a calibration chart to provide the required final concentration.

It is more convenient, however, for such an adjustment to take place automatically and most vaporisers contain a temperature controlled valve which adjusts the splitting ratio. The control systems illustrated in Fig. 11.8 are shown acting on the flow through the vaporisers. However, in some vaporisers they operate instead on the bypass channel. One temperature control valve incorporates a bimetallic strip as shown in Fig. 11.8A. This consists of two metals which have different coefficients of thermal expansion and which are joined together. As the temperature changes, the shape of the strip alters so that it bends or straightens. This strip may be used to operate a valve and is found in the Fluotec and PAC vaporisers and also in the Tekota and Cardiff inhalers. Another method, illustrated in Fig. 11.8B, is to use small flexible bellows containing some fluid that has a high coefficient of expansion. As the temperature changes, the bellows expand or contract and thus open or close a valve. This method is used in the EMO, Abingdon and Ohio vaporisers. In Dräger Vapor 19 vaporisers, expansion of a metal rod acts similarly to adjust the orifice to modify flow to the vaporising chamber according to the temperature (Fig. 11.8C).

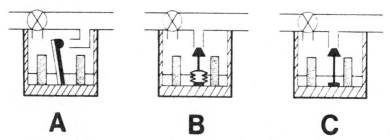

Figure 11.8 Temperature controlled valves to adjust automatically flow through the vaporising chamber based on: (A) a bimetallic strip, (B) bellows, and (C) a metal rod.

The problems arising from the fall of vapour pressure with temperature can be avoided if liquid anaesthetic is added directly to the gas stream, as in the Siemens vaporiser (Fig. 11.9). The anaesthetic agent is delivered into the gas stream through a fine nozzle. The rate of delivery of the anaesthetic depends upon the pressure difference P_1 to P_2 across the nozzle, and this is adjusted by the throttle valve. If flow through the vaporiser is increased, the pressure across the valve is increased and so more anaesthetic is delivered to maintain the same concentration. In this way the vaporiser remains accurate despite changes in flow, and the throttle valve is calibrated to indicate the percentage of anaesthetic delivered.

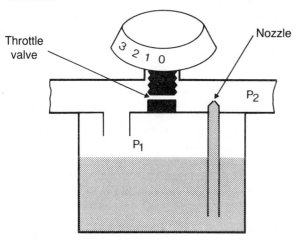

Figure 11.9   Siemens vaporiser.

EFFECT OF PRESSURE ON THE VAPORISER

The calibration of some vaporisers is inaccurate during intermittent positive pressure ventilation (IPPV), and the vaporiser may give increased concentrations of anaesthetic, particularly if low flow rates are used in association with IPPV with a closed circuit.

During IPPV, gas is supplied to the vaporiser at a constant rate which is set on the flowmeters on the anaesthetic machine, but intermittent back pressure from the ventilator can occur at the outlet of the vaporiser as the ventilator cycles. If the volume of the vaporising chamber in the vaporiser is larger than the bypass channel, gas may expand out of the inlet and outlet of the vaporising chamber when the back pressure from the ventilator falls. Thus, gas containing anaesthetic agent will flow into the bypass channel and increase the output concentration of the vaporiser.

This effect may be overcome in a number of ways. Firstly, a pressurising valve can be included downstream of the vaporiser to ensure that the pressure within the vaporiser is constant and greater then the pressure at the ventilator. Secondly, the vaporiser may be designed so that the volume of the bypass channel and vaporising chamber are equal. If this is the case, expansion takes place equally from both routes and there is no retrograde flow. Thirdly, a long inlet tube to the vaporising chamber can be provided so that retrograde flow from the vaporising chamber never reaches the bypass channel.

A further problem has occurred in vaporisers subjected to pressure, in that liquid anaesthetic has been decanted into the breathing system. This hazard is now avoided in the design of vaporisers.

USE AT HYPERBARIC PRESSURE

Anaesthetic vaporisers are also sometimes used at varying atmospheric pressure. Figure 11.10 reviews the situation with a vaporiser set at 1% halothane. If the ambient atmospheric pressure is 100 kPa (1 bar), the partial pressure of halothane at the outlet is 1% of this, i.e. 1 kPa. At room temperature, the saturated vapour pressure of halothane in the vaporiser is 32 kPa, or 32% of the atmospheric pressure.

Consider what happens if this same vaporiser with the same setting is operated in a pressure chamber at 200 kPa (2 bar). The saturated

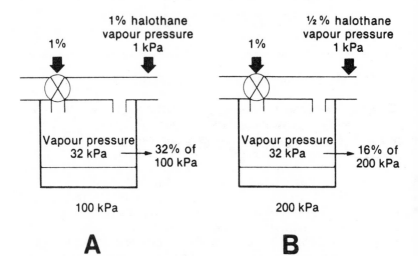

Figure 11.10   The effect of a hyperbaric pressure of 200 kPa on the performance of a halothane vaporiser. The percentage delivered is halved but the partial pressure is unchanged.

vapour pressure is unaffected by ambient pressure; hence it is still 32 kPa but this now constitutes 16% of the 200 kPa atmospheric pressure. The splitting ratio is unchanged and so, as well as a halving of the percentage of halothane inside the vaporiser, there is a halving of the output percentage measured at the ambient 200 kPa pressure. The vaporiser is thus delivering 0·5% halothane at 200 kPa which results in a partial pressure of halothane of 1 kPa, the same as before. As the depth of anaesthesia is dependent on the partial pressure of anaesthetic vapour rather than on its percentage, an anaesthetic vaporiser can normally be used with the usual settings at different atmospheric pressures.

VAPORISER POSITION AND CONTROLS

On the anaesthetic machine the vaporiser should be placed between the flowmeter block and the emergency oxygen flush control (Fig. 11.11) so that there is no risk that the high flow of oxygen from the latter can be delivered through it. If there is a gas supply cut-out actuated by a failure of the oxygen supply, then this too is downstream from the vaporiser.

The vaporiser control knob should be standardised to turn to 'off' in a clockwise direction as shown in Fig. 11.11. When the vaporiser is in the 'off' position both the inlet and outlet ports to the vaporising chamber should be occluded to avoid contamination of gas from the flowmeters with traces of anaesthetic.

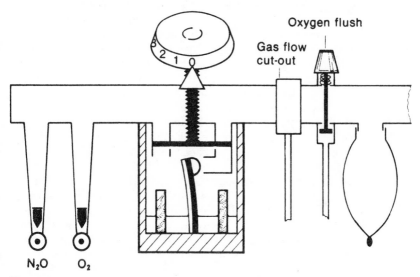

Figure 11.11 Aspects of vaporiser design and position.

USE OF VAPORISERS IN TANDEM

If two vaporisers are situated in tandem on the anaesthetic machine, the second one could become contaminated with vapour from the first. To solve this problem a simple vaporiser such as a Boyle's bottle may be placed in the second position because this can be readily cleaned and any contaminated anaesthetic discarded. Alternatively, as few anaesthetists wish to use two vaporisers simultaneously, some manufacturers provide an interlock mechanism to allow for rapid change of the vaporiser, if it is desired to change over to another agent.

DRAW-OVER VAPORISERS

The principles described for plenum vaporisers also apply to draw-over vaporisers, but the internal flow resistance of the draw-over vaporiser must be very low to avoid additional resistance to the patient's breathing. If such vaporisers are calibrated, then they must retain their accuracy over a wide range of tidal ventilation. An example of the difficulty in achieving this arose in the special draw-over vaporisers which were used to give methoxyflurane or trichloroethylene as analgesics in obstetrics. In labour, patients tend to hyperventilate and early models of these vaporisers did not remain accurate during use.

At present draw-over vaporisers are less popular for analgesia but those such as the EMO, the OMV (Oxford Miniature Vaporiser) and the PAC (Cyprane) are used in portable anaesthetic apparatus. Because they do not require cylinder gas supplies, they are ideal for use outside hospital for emergencies, in major disasters or in remote areas.

The EMO has a bellows thermal compensation device and is well-established. It is heavier than the other vaporisers mentioned as it also has a water reservoir to aid thermal stability. It is designed for use with ether and, as this depresses the respiration less than halothane, it is safer for use in circumstances when oxygen is not available. The EMO, however, is less suitable for use with a continuous flow, plenum system as it then gives a lower percentage of anaesthetic than indicated.

The OMV is conveniently compact and has a small water reservoir but no thermocompensating valve. It is designed for use with halothane, trichloroethylene or enflurane, preferably with a system of oxygen supplementation. It can be used with a continuous or an intermittent flow, or with a draw-over technique. In a development of the OMV, the Triservice vaporiser, the heat retaining reservoir contains antifreeze in place of water so that it is not damaged at low temperatures and continues to function in field use under winter conditions.

The PAC series of vaporisers have bimetallic strip temperature compensation, and their performance is not unduly affected by shaking, tilting or transient overturning—an advantage for vaporisers used in the field. Vaporisers are individually calibrated for the anaesthetic agent required and can be used in a plenum or draw-over system.

Draw-over vaporisers have also been used within an anaesthetic circle system using carbon dioxide absorption, but careful monitoring of the anaesthetic concentration is then needed as cumulative overdosage with the anaesthetic can occur, particularly if intermittent positive pressure ventilation is used. In addition, contamination of the vaporiser with water vapour may affect its performance.

CARE OF VAPORISERS

Most manufacturers recommend annual servicing of vaporisers, but the calibration of vaporisers can be checked with a refractometer if required (Chapter 13). In the case of halothane and some other anaesthetics it is recommended that the vaporiser is drained and refilled at regular intervals, usually weekly, to prevent accumulation of preservatives such as thymol in the vaporising chamber.

Care must be taken that calibrated vaporisers are filled only with the appropriate anaesthetic. To prevent accidental filling of vaporisers with the wrong liquid, a safety system is available using filler tubes and caps which only fit the appropriate vaporisers and bottles of anaesthetic.

12

Humidification

Anaesthetists' main interest in humidification is in patients, in particular those whose normal humidifying mechanism in the nose and mouth has been bypassed, perhaps by tracheostomy or endotracheal intubation.

In addition, humidity is of importance in the operating theatre because high humidity is unpleasant while low humidity may allow the build-up of static charges with a risk of explosion if flammable agents are in use.

UNITS OF HUMIDITY

Humidity may be expressed in two ways. The first is the absolute humidity.

  Absolute humidity is the mass of water vapour present in a given volume of air.

The values of absolute humidity are usually expressed as mg litre^{-1} or g m^{-3}, and these two measurements are numerically the same. The amount of water vapour that can be present in a given volume of air is limited by the temperature.

As illustrated in Fig. 12.1, when temperature increases, the amount of water which can be present as vapour also increases. Hence fully saturated air at 20 °C contains about 17 g m^{-3}, whereas at 37 °C it contains 44 g m^{-3} when saturated.

The second way of measuring humidity (relative humidity) is a comparative method.

  Relative humidity is the ratio of the mass of water vapour in a given volume of air to the mass required to saturate that given volume of air at the same temperature. It is usually expressed as a percentage.

From the graph (Fig. 12.1) it may be seen that 1 m^3 of air at 20 °C, 100% saturated, contains about 17 g water. If it is warmed to 37 °C, the mass

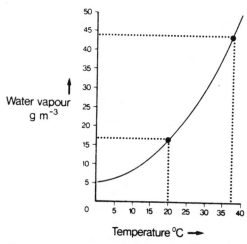

Figure 12.1 Graph of the humidity of air, saturated at various temperatures.

of water vapour or absolute humidity is still the same but the relative humidity is only 39%, as at 37 °C 1 m³ of air contains 44 g water vapour when fully saturated and the ratio of 17 to 44 gives a value of 39%.

Consider again the definition of relative humidity.

$$\text{Relative humidity} = \frac{\text{Mass present, } M_p}{\text{Mass to saturate, } M_s}$$

From the gas laws (Chapter 4):

$$n = \frac{PV}{RT}$$

and mass

$$M \propto n$$

where n = number of moles
   P = absolute pressure
   V = volume
   R = universal gas constant
   T = absolute temperature

Therefore:

$$\text{Relative humidity} = \frac{\dfrac{P_p V}{RT}}{\dfrac{P_s V}{RT}}$$

Therefore, if V, R and T are constant then:

$$\text{Relative humidity} = \frac{P_p}{P_s}$$

$$= \frac{\text{Actual vapour pressure}}{\text{Saturated vapour pressure}}$$

It is seen that relative humidity can also be expressed as a ratio of vapour pressures.

MEASUREMENT OF HUMIDITY

Most instruments measure humidity as relative humidity.

Figure 12.2 shows a hair hygrometer. It may be mounted on a wall of the operating theatre and gives a direct reading of relative humidity.

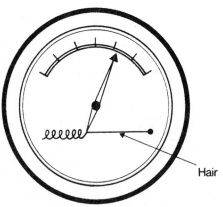

Hair

Figure 12.2   Hair hygrometer.

The instrument uses the fact that a hair gets longer as the humidity rises, and the hair length controls a pointer moving over a scale. It has the advantage of simplicity, and is most accurate for humidities between 30 and 90%. Another form of meter works on a similar principle but has a special membrane in place of the hair.

Another instrument used for relative humidity measurement is the wet and dry bulb hygrometer (Fig. 12.3). It consists of two thermometers, as shown. The temperature of the mercury in the bulb on the right is in equilibrium with its surroundings and reads the true ambient temperature. However, the left thermometer reads a lower temperature because of the cooling effect from the evaporation of water from the

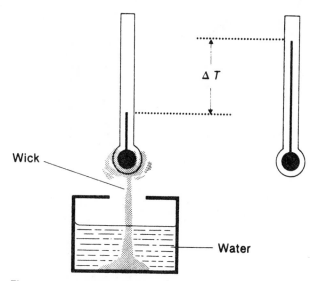

Figure 12.3 Wet and dry bulb hygrometer.

wick surrounding its bulb and the consequent loss of latent heat of vaporisation. The difference between the temperature registered by the two thermometers, ΔT, is related to the rate of evaporation of water which, in turn, depends on the ambient humidity. By using a set of tables it is possible to obtain relative humidity from this temperature difference. Notice that there must be sufficient air movement in the vicinity of the wet bulb to prevent a localised enhancement of humidity, and accurate instruments supply a controlled airflow over the thermometers.

A more accurate instrument for humidity measurement is Regnault's hygrometer (Fig. 12.4). It consists of a silver tube containing ether. To use the hygrometer, air is blown through the ether, cooling it and so initiating condensation on the shiny outside surface of the tube. The temperature at which condensation or misting occurs can then be noted and is known as the dew point, the dew point representing the temperature at which the ambient air is fully saturated. Relative humidity may be deduced from the dew point as follows:

$$\text{Relative humidity} = \frac{\text{Actual vapour pressure}}{\text{s.v.p. at that temperature}}$$

$$= \frac{\text{s.v.p. at dew point}}{\text{s.v.p. at ambient temperature}}$$

where s.v.p. = saturated vapour pressure

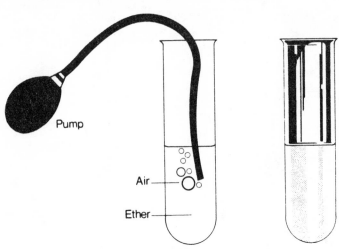

Figure 12.4   Regnault's hygrometer.

From a knowledge of the dew point and the saturated vapour pressures, which can be obtained from tables, both the relative and absolute humidities can be calculated at the temperature required.

Absolute humidity can also be measured by transducers which are of two main types depending on the change either in electrical resistance or in capacitance of a substance when it absorbs water vapour from the atmosphere. The mass spectrometer (Chapter 20) has also been used in humidity studies and humidity can also be measured by a light absorption technique based upon the reduction of the ultra-violet light transmitted when water vapour is present.

CLINICAL ASPECTS OF HUMIDITY

When the patient breathes through his nose the inspired air is warmed and saturated with water vapour before entering the trachea. If the nose is bypassed by an endotracheal tube or tracheostomy, dry air enters the trachea. The secretions present in the trachea may therefore become dried and tenacious and mucous plugs may form that block the respiratory tract. The secretions themselves are then more difficult for the patient to cough up or for the anaesthetist to aspirate. The cilia too may be inhibited or damaged by the dry gases, which also tend to have a chilling effect due to the latent heat lost from the mucosa as moisture evaporates to humidify the dry inspired gases. This inhibits ciliary activity further. If cilia are exposed to dry gases for a long period, they disappear and the epithelium of the trachea becomes keratinised. Such effects could occur in patients on long-term intensive care but humidification has also been advocated for the inhaled gases in routine anaesthesia.

Normally, the air entering the upper trachea is saturated with water vapour to a level of 34 g m^{-3}, i.e. it is almost fully saturated at a temperature of 34 °C. This provides a suitable level of humidity as a standard when humidifying inspired air in patients whose upper respiratory tract is bypassed, e.g. by a tracheostomy.

There are two methods of artificially increasing the inspired humidity. The first is by humidifying the environment, and the second is by humidifying the inspired gases alone. Increases in total environmental humidity occur in oxygen tents and infant incubators. In anaesthesia, however, humidifiers are commonly used to humidify the inspired gases alone.

HUMIDIFIERS

The simplest method of humidifying the inspired gases is the instillation of water droplets directly into the trachea, but this is the least satisfactory technique and cannot be recommended for routine use. It carries the danger that, if the drip is run too rapidly, the water or saline instilled into the trachea may lead to a condition similar to the acid-aspiration syndrome.

Another type, the condenser humidifier, also has the advantage of simplicity and it is sometimes called the artificial nose.

A typical variety (Fig. 12.5) consists of an inlet and outlet which can

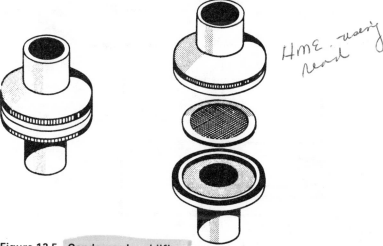

HME: using read

Figure 12.5   Condenser humidifier.

be screwed together with a mesh, sometimes in the form of a metal gauze, sandwiched between them. The humidifier conserves the water which is normally lost on expiration. When the warm, moist expired gases pass through the metal mesh they become cooled and the water condenses, only to be evaporated again on the next inspiration. The efficiency of these condensers depends on the prevailing temperature of the inspired air; thus if the humidifier were used in a hot tropical climate, there would be little temperature change and little water would condense out on the metal gauze. Under optimum conditions, however, these small humidifiers can provide adequate humidification. A problem is that hardened secretions may be deposited on the metal gauze, causing a marked rise in resistance to breathing. For this reason, and also to prevent infection, the gauze should be changed every few hours or as advised by the manufacturers. Even so, there is still a risk of the gauze acting as a prolific source of infection with organisms such as pseudomonas. One form of condenser humidifier incorporates a hygroscopic unit in addition to the usual condenser unit and so achieves greater efficiency.

Another system of humidification uses a water bath; in its simplest form this system consists of dry gases bubbling through water at room temperature, but there are then two main disadvantages. Firstly, such a system lacks efficiency because of the large bubbles, although by passing the gas through sintered glass to reduce the bubble size, the efficiency can be increased. The second problem is the loss of heat from the water by latent heat of vaporisation resulting in cooling which

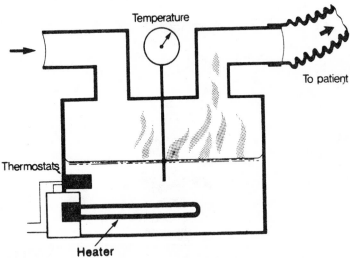

Figure 12.6 Hot water bath humidifier.

reduces the humidity obtainable. This problem of cooling is overcome by adding an electric heater (Fig. 12.6), and adjustment of the water bath temperature gives flexibility in the amount of water vapour which can be added to the gas stream.

The temperature at which the humidifier is operated depends on the humidity required at the patient and the temperature drop along the tubing. The operating temperature is normally about 40 to 45 °C, but higher temperatures up to 60 °C may sometimes be employed to prevent the growth of bacteria within the humidifier. There is a danger of scalding the patient if too high a temperature is present. This risk may be further increased in the case of patients on ventilators, especially if a ventilator fault occurs that results in a sudden increase in flow through the humidifier.

If the patient continuously inspires fully saturated gas at body temperature or higher, the heat normally lost via the respiratory tract in the warming and humidification of the inspired air is prevented. In fact, heat may even be added to the patient and, if other physiological methods of heat loss are depressed and inadequate, hyperthermia may occur especially in young children.

To avoid the dangers of overheating of the humidifier, a thermostat, usually operated by a bimetallic strip, is present (Fig. 12.6). If the water overheats, the thermostat automatically switches off power to the heater. In modern humidifiers there is usually also a second thermostat in reserve in case the first fails as well as a thermometer to indicate the temperature of the humidifier.

A potential problem with this humidifier is illustrated in Fig. 12.7. The water vaporised can condense in a redundant loop of the delivery

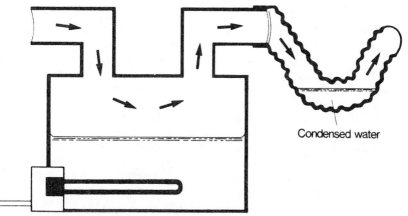

Condensed water

Figure 12.7 Condensation of water in a loop of delivery tube from a hot water bath humidifier.

tubing to obstruct the gas flow to the patient; such a pool of hot water could even be blown into a patient's tracheostomy. It is important, therefore, to ensure that the water condensing out can run back to the humidifier or into a water trap.

Figure 12.8 illustrates further refinements to the heated water bath principle. This is the cascade humidifier, in which gas bubbles through a perforated screen at the bottom of a wide tube causing a foam of water and gas to pass up the tube. The large surface area of gas exposed to the

water bath humidifier

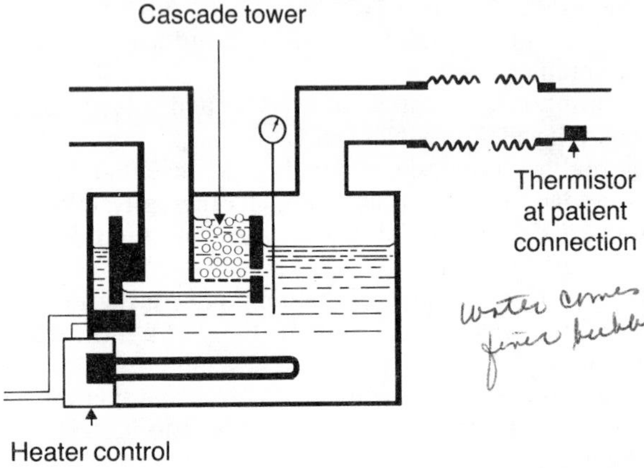

Cascade tower

Thermistor
at patient
connection

Water comes out in finer bubbles.

Heater control

Figure 12.8   Cascade humidifier.

water ensures that it is fully humidified. The thermistor positioned at the patient connection on the right of the figure monitors the temperature of the gases delivered and controls the heater element in the water bath to maintain an optimal delivery temperature and avoid any risk of burns to the patient. This gives safer and more accurate control than is possible when only the water bath temperature is monitored. As the gases are fully saturated, the temperature at the patient also indicates absolute humidity at this point.

An alternative technique for adding water as vapour is the heated element humidifier illustrated in Fig. 12.9. Water is vaporised by dripping it on to an electric element heated to 100 °C, the high temperature ensuring sterility. Also shown is a water trap which collects excess water. In an alternative form of this device, small pulsed volumes of water are delivered to the heated element and an element temperature of 250 °C is used. However, humidifiers with such hot elements are less suited to use with anaesthetic vapours or volatile drugs as the heat may cause chemical changes. The amount of water vapour delivered

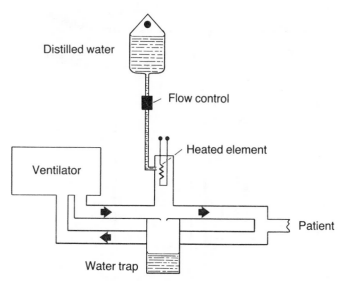

Figure 12.9 Heated element humidifier.

from these humidifiers must be controlled according to the minute volume and humidity required, and a thermistor at the patient connector may be present to assist this control and avoid any risk of burns.

NEBULISERS

The nebuliser is another form of humidifier and may be of two types, gas-driven or ultrasonic.

The gas-driven nebuliser depends on the Bernoulli effect (Chapter 2). A flow of high pressure gas across an orifice causes a drop in pressure which allows water to be drawn up or entrained from the side tube (Fig. 12.10). The entrained water is then broken up into a spray of droplets of various sizes in the high-pressure gas stream. The larger droplets can be baffled out and more finely divided if they are allowed to hit an anvil as shown.

Because some of the finer droplets tend to vaporise, there is a fall in temperature in the humidifier, so its efficiency can be increased by adding a heater. A problem of this type of nebuliser is that back pressure may alter the entrainment ratio, and can therefore affect the efficiency of the device.

In the ultrasonic form of nebuliser the droplets of water are formed by a vibrating surface (Fig. 12.11). The water may be dropped

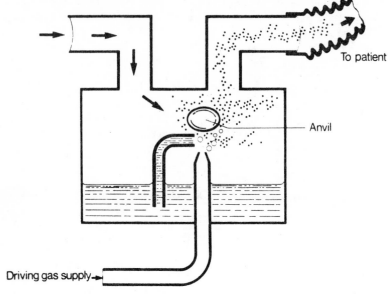

Figure 12.10   Gas-driven nebuliser.

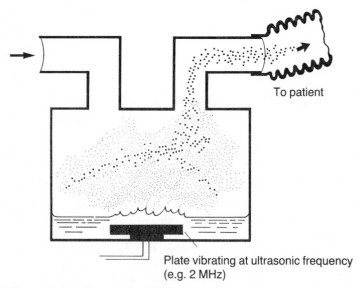

Figure 12.11   Ultrasonic nebuliser.

on a vibrating surface or formed from a pool lying on such a surface. The frequency of vibration is usually a few megahertz which produces extremely small droplets.

Size of Droplets

The size of the droplets produced by a nebuliser is important. It has been shown that droplets of over 20 microns (μm), if not baffled out, are a nuisance as they fall out to form pools of water either in the tubing or in the upper respiratory tract. Those of 5 μm fall out in the region of the trachea, while any particles of 1 μm tend to pass right through to the alveoli and be deposited there. Droplets below 1 μm in size tend to be extremely stable and can be inspired and then expired again. One micron is ideal and ultrasonic nebulisers are most efficient at producing this size of water droplet, but slightly larger particle sizes may be helpful to loosen secretions in the trachea.

By the addition of droplets to the inspired gas, it is relatively easy to obtain moisture levels corresponding to relative humidities of well over 100%, and so, especially with children, it is easy to overload the patient with water. In addition to this danger of water intoxication, problems may arise from hypoxia if too high a humidity is provided. This hypoxia can be caused by shunting of blood with alteration of ventilation–perfusion ratios due to non-functioning alveoli. Hypoxia may also be caused by an increased resistance to breathing. This may arise from narrowing of the diameter of the air passages by the film of water or wave formation in this film. The inspired droplets also increase the density of the gases and raise the resistance to turbulent flow.

One further point to note is that as small droplets are stable and can carry for long distances they may be a prolific source of cross-infection, particularly if a humidifier becomes infected. Solutions used in the humidifier must be sterile and the humidifier kept free from contamination.

RELATIVE EFFICIENCY OF HUMIDIFIERS

A diagram of the approximate efficiencies of the various humidifiers is illustrated in Fig. 12.12, in which absolute humidity readings are used to avoid the need to specify the temperature. There are vertical dotted lines at 34 g m^{-3} representing the normal level of humidity in the upper trachea and at 44 g m^{-3} representing full saturation at 37 °C.

The unheated humidifier is not very efficient. The condenser humidifier is more efficient, but the gauze must be changed frequently. The heated-water bath humidifier and the heated Bernoulli type of nebuliser provide humidity close to that normally present in the upper trachea. The ultrasonic nebuliser is the most efficient method of adding water to the inspired gases, but it is expensive. Its efficiency also leads to the need for special care to avoid overdosage.

Finally the problem of cross-infection must always be remembered, regardless of the type of humidifier used.

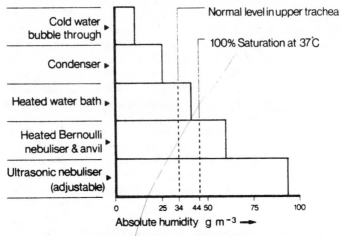

Figure 12.12 Approximate efficiency of humidifiers. The exact values depend on the model of humidifier in use.

The Sine Wave and Wave Patterns

THE CONCEPT OF A SINE WAVE

In medicine there are various biological processes that occur in a repetitive manner, e.g. the ventilation of the lungs and the cardiac cycle. When these are monitored, a complex wave pattern is seen. Figure 13.1 shows tracings of arterial pressure, of the electrocardiogram (ECG), and of the electroencephalogram (EEG) and also gives examples of different types of simple wave forms—i.e. the sine wave, the sawtooth wave, and the square wave. Of these, the sine wave is of special importance because it is possible to produce any of the other patterns by combining together various different sine waves.

Wave motion is also present as variation in pressure in the case of sound waves and variation in the electrical and magnetic fields in the case of electromagnetic radiation such as light.

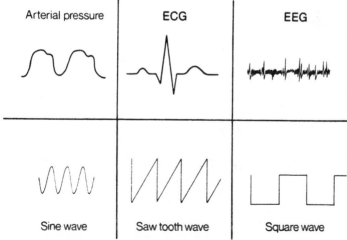

Figure 13.1 Wave patterns.

Two illustrations of sine wave production are given in Fig. 13.2. On the right, the movement of the coil in an electrical generator produces a sine wave pattern of electric current. This type is known as alternating current (a.c.) electricity, and is discussed in Chapter 14. Sine waves may also be produced mechanically, and in both cases the principle involved is the same. Illustrated on the left of Fig. 13.2 is a pump with a piston

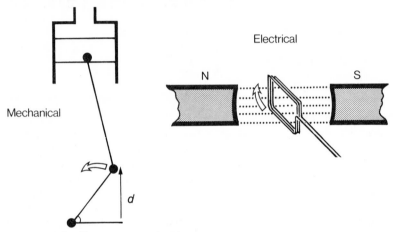

Figure 13.2   Sine wave production.

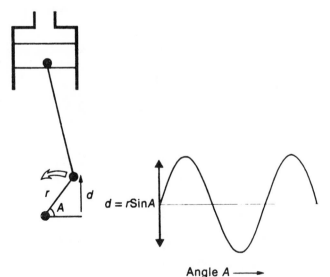

$$d = r\,SinA$$

Angle $A \longrightarrow$

Figure 13.3   The sine wave.

moving up and down with the movement of a crankshaft. At the bottom of the diagram, the rotation of the crankshaft produces a vertical displacement d, and if the vertical movement of this crankshaft is traced against time a sine wave is produced.

As shown in Fig. 13.3, the vertical displacement d and the angle of the crankshaft A form a right-angled triangle with the crankshaft of fixed length r as the hypotenuse. The sine of the angle A is equal to d divided by r so, regardless of the position of the crank at any time, the distance d is equivalent to r sine A. Angle A has a different value at each moment because the crank is rotating at a constant rate and so a graph can be obtained with the displacement d on the vertical axis and the angles of A corresponding to the different times along the horizontal axis. This results in a sine wave pattern as shown on the right.

CHARACTERISTICS OF THE SINE WAVE

Phase

The different angles of A along the horizontal axis are referred to as the phase of the sine wave. One complete rotation of the pump crankshaft results in one complete cycle on the graph, and so 360° corresponds to one complete cycle (Fig. 13.4). A rotation of the crank through 180° is half a cycle, through 90° a quarter of a cycle.

If two sine wave motions are compared as shown on the graph (Fig. 13.5), the relative displacement of one wave with respect to the other can be referred to in terms of the number of degrees by which the two

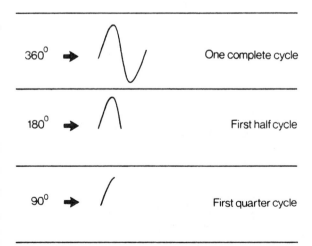

Figure 13.4   Phase of a sine wave.

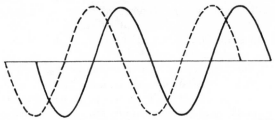

Figure 13.5   The two sine waves illustrated are 90° out of phase.

waves differ along the horizontal axis. In the example illustrated, the two sine waves differ by exactly one-quarter of a cycle or 90°, and so they are said to be 90° out of phase.

Wavelength and Amplitude

Consider now the sine wave shown in Fig. 13.6. The wavelength is the distance between any two corresponding points in successive cycles—i.e. the distance between two peaks or two troughs, or the distance between two points where the wave motion cuts the horizontal axis in the same direction. The maximum displacement of the wave from the horizontal axis is known as its amplitude, and is also marked on the illustration. In the case of sound waves, a large amplitude is loud and a small amplitude quiet, whereas with light waves a large amplitude is bright and a small amplitude dim.

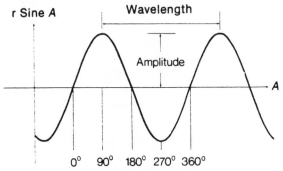

Figure 13.6   Wavelength and amplitude.

Frequency, Period and Velocity

The frequency of a wave is the name given to the number of complete cycles which occur in 1 second, and in the SI system the unit of measurement for cycles per second is given the name hertz and is denoted by the symbol Hz.

In Fig. 13.7 the wave form has a frequency of 10 cycles per second or 10 hertz and, as ten complete cycles occur in 1 second, the time taken for each cycle is one-tenth of a second. The time taken for one complete cycle to occur is referred to as the period of the wave motion and is the reciprocal of the frequency.

$$T = \frac{1}{f}$$

where T = period

f = frequency

Consider now the velocity of a wave motion:

Velocity = Frequency × Wavelength

Figure 13.7 illustrates a wave motion with a frequency of 10 Hz. If it has a wavelength of 1 m, the speed of the wave motion is then 10 m s^{-1}.

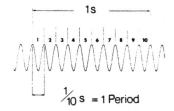

Frequency 10 Hz = 10 Cycles per second

Figure 13.7   Period and frequency.

Different wave motions have different velocities; sound, for example, has a velocity of 330 m s^{-1} in air. If the velocity is fixed, then frequency and wavelength are inter-related, the higher the frequency, the shorter the wavelength and vice versa.

Sound waves of different frequencies are picked up by the ear as changes in pitch—for example, a sound wave with a high frequency has a short wavelength and a high pitched note, whereas a sound with a low frequency has a longer wavelength and sounds low pitched.

The behaviour of light can also be explained in terms of wave motions. A light wave motion with a high frequency and a short wavelength is seen as blue, whereas one with a lower frequency and longer wavelength appears red.

Light is part of the larger electromagnetic spectrum, illustrated in Fig. 13.8. Electromagnetic radiation may be described by its characteristic frequency. For example, radio waves have frequencies up to 10^9 Hz, whereas x-rays and gamma-rays have frequencies over 10^{21} Hz. The

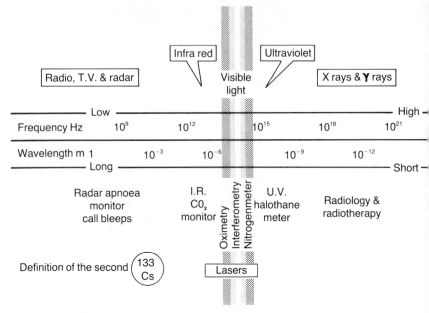

Figure 13.8 Electromagnetic spectrum.

corresponding range of wavelengths is also shown on the diagram. Because the velocity of all the wave motions of the spectrum is constant at 3×10^8 m s^{-1}, the wavelength is related to the reciprocal of the frequency. The wavelengths of the spectrum thus vary from 1 m to 10^{-12} m, and each type of radiation has a characteristic wavelength associated with it.

Visible light is just a tiny part of this larger spectrum and consists in itself of a spectrum of colours. Radiations whose wavelengths are just outside those of the visible range are called infra-red and ultra-violet.

Use is made of the different electromagnetic waves in medicine. The 'bleep' system used to call staff, and a radar apnoea monitor use radio waves. Gases are capable of absorbing electromagnetic radiation and use has been made of this fact to produce infra-red meters for carbon dioxide and nitrous oxide analysis, and an ultra-violet meter to measure halothane concentrations. Infra-red rays are also used in thermography, and ultra-violet light is used in certain recorders. Gamma-rays and x-rays are essential to radiology and are used in radiotherapy and nuclear medicine. Visible light is also used in many medical instruments, e.g. in the oximeter used to measure oxygen saturation in blood (Chapter 18) and in interferometry, considered later. Lasers (Chapter 24) use light in the visible and infra-red wavelengths.

Gases emit electromagnetic radiation if they are suitably excited, e.g. by bombardment by electrons, and the wavelengths of radiation emitted are characteristic of the gas. In the SI system the second is defined in terms of the frequency of radiation emitted by atoms of caesium-133. Finally, the radiation emitted by nitrogen when excited forms the basis of a nitrogen meter.

THE DOPPLER EFFECT

An example of the Doppler effect is the change of the pitch of sound emitted from a rapidly moving vehicle as it passes the observer. As mentioned above, if the pitch of a note changes, a change in frequency must have occurred. Consider now how this happens. Sound waves are regions of high and low pressure in the air, and they travel through it at a fixed velocity. At the top of Fig. 13.9 is a stationary source of sound emitting sound waves of a particular frequency. The separation between high and low pressure regions depends on the wavelength and therefore on the frequency. Now suppose the source, still emitting the same frequency, starts moving towards the listener. In the interval between producing high pressure regions, the source has moved forward and so each high pressure region becomes closer to the previous one. Consequently, the wavelength of this sound becomes shorter. Because of the relationship between wavelength and frequency the ear of the listener picks up this shorter wavelength as an increase in frequency compared with the same source when stationary and the

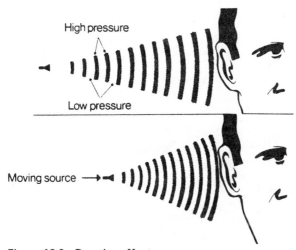

Figure 13.9 Doppler effect.

pitch of the note is higher. Conversely, as a source moves away from a listener, he hears a lower pitch from the drop in frequency. These changes in frequency constitute the Doppler effect.

Various uses are made of the Doppler effect in medicine. One application is the ultrasonic blood flow detector (Fig. 13.10). Ultrasound is very high frequency sound which is above the range of human hearing but which travels well through body tissues. In this apparatus ultrasonic waves from a vibrating crystal transmitter are beamed along an artery and the red blood cells reflect these high frequency sound waves. A receiving transducer, incorporated into the pencil-like head of the apparatus, detects the reflected sound waves and, because of the movement of the red blood cells, these reflected sound waves have a Doppler change in frequency. This change in frequency is sensed electronically and is related to the velocity of the blood cells. Although it is difficult to calibrate the apparatus to provide quantitative measurements of blood flow, this technique can detect blood flow in an artery and is used to assess qualitatively the degree of success of arterial grafts, especially in the leg. The human ear cannot detect this high frequency ultrasound but the apparatus often incorporates a system to produce audible sound related to the flow velocity. If the detector is used with a recorder, analysis of the tracing gives additional information regarding the flow pattern through the artery. A similar type of apparatus using the same principle has been designed to assess blood flow in other vessels—for example, an indication of cardiac output may

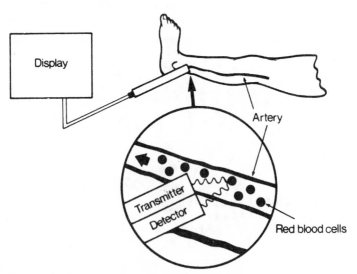

Figure 13.10 Ultrasonic blood flow detector.

be obtained if a detector is used to monitor the velocity of the blood in the arch of the aorta.

Another application of the Doppler shift which also uses ultrasonic waves is the ultrasonic blood pressure detector which is designed and positioned to detect arterial wall movement in preference to blood cell movement. This instrument is considered in Chapter 17.

A similar type of apparatus is used in obstetrics to monitor the fetal heart rate in labour (Fig. 13.11). In this case the apparatus picks up the ultrasonic beam as it is reflected off the moving fetal heart. The apparatus consists of a transmitter crystal and a receiving transducer which are placed on the abdominal wall. Using this apparatus any change in the fetal heart rate during labour is detected more rapidly and more accurately than by the usual method of auscultation. In addition, the apparatus may be used to detect and measure the fetal heart rate in early pregnancy.

Ultrasonic waves are also used to form images of body structures because the waves are reflected off boundaries and interfaces between substances of different densities. In this case it is not the Doppler effect, but the time taken for the ultrasonic wave to travel from the transmitter to the receiver which allows the depth of a boundary or change in density to be measured. By moving the transmitter a picture of a cross-section of the body can be built up. Examples of such ultrasonic scans are found in obstetrics and gynaecology, cardiology (known as echocardiograms) and urology. In this case the same crystal is used to transmit and receive the ultrasound waves.

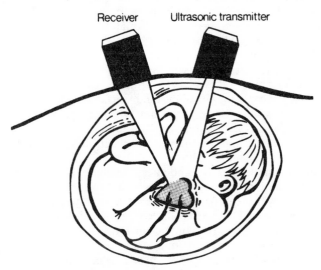

Receiver   Ultrasonic transmitter

Figure 13.11   Fetal heart rate monitor.

INTERFEROMETRY

Consider what happens when two or more sine waves are added together. Figure 13.12 shows on the left the result of adding two sine waves which have the same frequency, amplitude, and phase. The resultant sine wave has a greatly enhanced amplitude. On the other hand, if these two sine waves are 180° out of phase, the resultant sine wave has zero amplitude and appears as a straight line, as shown on the right of the figure.

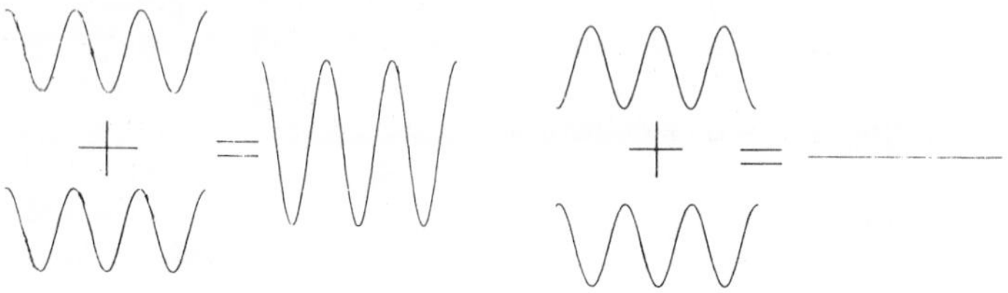

Figure 13.12 Addition of sine waves of same frequency and amplitude.

This interaction of two such sine waves is called interference and the resultant wave depends on the phase of the two wave motions which interfere. This phenomenon is used in optical devices for analysing various gases or vapours. An example is the interferometer, sometimes called a refractometer, with which halothane concentrations in oxygen may be measured and the performance of vaporisers checked. Two light beams pass through the refractometer, one through a sample chamber and one through a reference chamber (Fig. 13.13).

Consider what happens when these two beams are now added together: either they arrive at a point with their wave motions in phase, in which case reinforcement takes place and the increased amplitude gives rise to a bright fringe; or they combine out of phase and the reduced amplitude gives rise to a dark fringe. Thus, as shown on the right of Fig. 13.13, an array of light and dark fringes results. Suppose the reference chamber is left unchanged and an unknown concentration of sample gas added to the sample chamber. This reduces the velocity of the light passing through the sample chamber and alters the phase relationships when the two light beams are recombined, and so a displacement of the fringe pattern occurs which can be observed through an eye piece. The displacement of a selected fringe is directly proportional to the concentration of gas added to the sample chamber.

Small portable interferometers are commonly used in anaesthesia, but more sensitive versions also exist. The refractometer is non-specific and may be used for measuring the concentration of a variety of

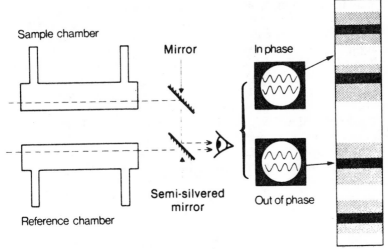

Figure 13.13   Principle of the interferometer.

different gases. On the other hand, it cannot be used to identify component gases in a mixture and so these must be known in advance.

For example, when a refractometer is used to check the performance of vaporisers, the refractometer calibration using a halothane–oxygen mixture is not identical with that for a halothane in nitrous oxide and oxygen mixture.

THE FREQUENCY SPECTRUM

Figure 13.14 illustrates the results of adding sine waves with differing frequencies and amplitudes, as well as a differing phase relationship. It

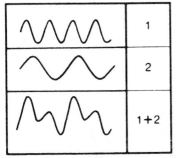

Figure 13.14   Addition of sine waves of differing frequencies and amplitudes.

is seen that the resultant waveform is no longer a sine wave but in this case is a wave pattern similar to an arterial pressure tracing. This result is obtained simply by adding the amplitudes of the component waves at every point along their cycles.

The addition of whole ranges of sine waves, each with a different frequency, may result in quite complex waveforms and, as an example, Fig. 13.15 shows how an ECG waveform could be produced by adding a large range of frequencies. The range of these frequencies is important in the design and use of monitoring apparatus and may be described by a spectrum of frequencies.

In such a spectrum (Fig. 13.16) the ECG is composed of a range of frequencies from about 0·5 to 80 Hz, while an EEG signal has a

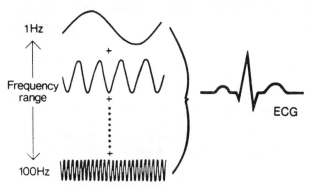

Figure 13.15 Component frequencies of the ECG.

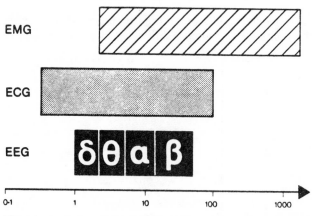

Figure 13.16 Frequency distribution of three biological electrical signals.

smaller range of frequencies. The slower waves of the EEG are only about 1 to 5 Hz, whereas the faster beta waves are in the range of 15 to 60 Hz. In the EMG (electromyogram) there are sharp spikes and these indicate high frequency components as shown.

Any waveform, no matter how complex, can be produced by selecting and adding together appropriate sine waves. The mathematical process of analysing complex wave patterns into a series of simpler sine wave patterns is called Fourier analysis, and this concept helps in understanding the patterns of biological electrical signals. The two main points to note are that, firstly, any complex wave pattern may be analysed into a large number of sine wave components and, secondly, wave patterns that have sharp spikes in them have high frequency components, while smooth and rounded wave patterns consist of a more limited range of frequencies.

14

Electricity

STATIC ELECTRICITY AND ELECTRICAL POTENTIAL

The word electricity is derived from 'ēlektron', the Greek word for amber. As with all substances, the atoms in amber consist of a positively charged nucleus surrounded by negatively charged electrons. Rubbing amber against another material can lead to a transfer of electrons, so that one of the substances will have an excess of them and the other a deficit. This phenomenon may also occur with other substances—e.g. plastics and rubber—and is known as static electricity. Such static electricity can be a source of ignition in anaesthetic explosions (Chapter 23). Static electricity is not the only cause of one object having an excess of electrons and another a deficit. This phenomenon also occurs as a result of chemical reactions in batteries, and in biological tissues. The electrons in the objects with the excess then possess potential energy in the same manner that the height of an object determines its gravitational potential energy. There is said to be an electrical potential difference between the object with the excess and that with the deficit.

CONDUCTORS, INSULATORS AND SEMICONDUCTORS

Under the influence of a potential difference (or a magnetic field, dealt with below) electrons can pass from one atom to another and an electric current is the movement of such electrons through a substance. Substances may be classified into conductors, insulators or semiconductors according to their ability to conduct electrons.

Conductors are usually metals, although certain other substances such as carbon are also good conductors. The outer electrons of atoms of a metal are loosely bound and can readily move through the metal under the influence of an electric potential.

In insulators the electrons are firmly bound and therefore not normally able to move and form an electric current.

Semiconductors are used in components such as thermistors, transistors and diodes. The conductivity of semiconductors is intermediate between that of conductors and insulators and is usually strongly dependent on temperature; thus thermistors are used for temperature measurement (Chapter 9). Because anaesthetic monitoring equipment contains semiconductors, excessive heat adversely affects performance.

MAGNETIC FIELD *produced by electricity*

It is found that a conductor with a current flowing through it can exert a force on another conductor carrying a current; this is the property known as magnetism. Some substances, e.g. iron, can exhibit magnetism although it appears that no current is flowing through them. In this case, the magnetism is due to the sum of the many minute currents formed by the motion of electrons orbiting their nuclei.

The region throughout which a magnet or a current-carrying conductor exerts its effects is known as a magnetic field, and a changing magnetic field induces a flow of electrons in a conductor to produce an electric current.

DIRECT AND ALTERNATING CURRENT

The term direct current (d.c.) describes a steady flow of electrons along a wire or through a component in one direction only (Fig. 14.1). A common source of direct current is a battery, but there are other sources such as a thermocouple (Chapter 9).

The term alternating current (a.c.) describes a flow of electrons first in one direction and then in the opposite direction along a wire. A graph of alternating current against time may appear as a sine wave although

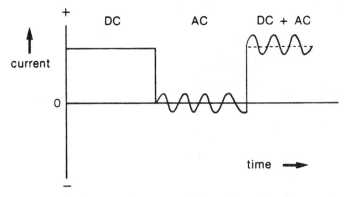

Figure 14.1 Graph of current against time for direct current, alternating current and a combined current form.

more complex waveforms, as described in Chapter 13, are also possible. A current which does not have a steady value but in which the electrons move on average in one direction may be described in terms of an a.c. component added to a d.c. component, as shown on the right of Fig. 14.1.

THE AMPERE AND CURRENT MEASUREMENT

The ampere (A) is the unit of current in the SI system. It represents a flow of 6.24×10^{18} electrons per second past some point, and is defined by means of the electromagnetic force which is associated with an electric current. As mentioned above, whenever an electric current flows in a wire, an associated magnetic field exists around the wire. Moreover, whenever a wire carrying an electric current is placed in a magnetic field there is a force which tends to move it in a direction perpendicular to both the electric current and the magnetic field.

The interaction of an electric current and a magnetic field is the working principle of instruments such as the galvanometer. In a galvanometer (Fig. 14.2) a coil of wire is suspended on jewelled bearings in a magnetic field. The current to be measured passes through this coil and the interaction between the electric current and the magnetic field

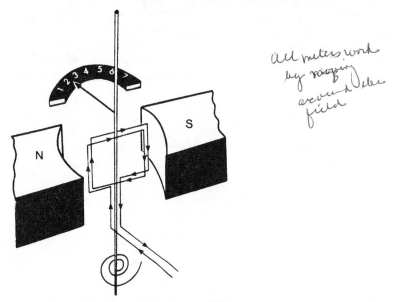

Figure 14.2   Principle of the galvanometer.

causes the coil to rotate. The rotational force on the coil is balanced by a hair spring, shown at the bottom of the diagram. The deflection of the coil is proportional to the electric current passing through it and is indicated by a pointer which moves over a scale.

Many display devices and recorders used in anaesthetic equipment are based on the principle of the galvanometer.

THE ELECTROMAGNETIC FLOWMETER

The electromagnetic flowmeter is another example of the application of electromagnetic effects. If a conductor is moved through a magnetic field, an electric potential develops and the magnitude of this induced potential is proportional to the rate at which the conductor is moved through the magnetic field. As blood is a good conductor of electricity, this principle may be used to measure blood flow.

Consider the artery shown in Fig. 14.3 which lies in a magnetic field produced by the current-carrying coils A and B, the direction of the field being perpendicular to the flow of blood. The blood forms a moving conductor and the potential developed as it flows through the magnetic field is perpendicular to both the direction of flow and the

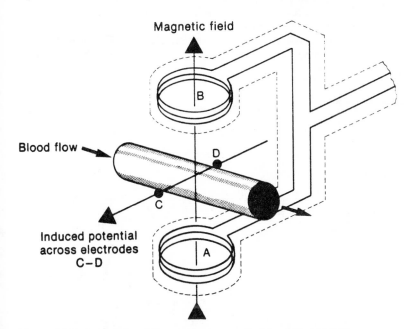

Figure 14.3   Principle of the electromagnetic blood flowmeter.

magnetic field, i.e. it is developed across opposite walls of the artery. This potential can be measured by two electrodes, C and D, touching opposite sides of the artery.

An electromagnetic flowmeter consists of a C-shaped probe, which contains coils to generate the magnetic field, and two electrodes to measure the potential induced as the blood flows. Although the potential is proportional to the rate at which the blood is flowing, the velocity of the blood varies across the diameter of the blood vessel and so the electromagnetic flowmeter measures an average velocity. It is important that the flowmeter is calibrated with the same type of flow as that which will be present in the vessel on which the measurements are to be made.

In practice, an alternating rather than a steady magnetic field is used and the alternating component of the potential is measured. This improves the stability of the measured value.

CURRENT FLOW, ENERGY PRODUCTION AND THE VOLT

When a current flows, heat and sometimes light energy is produced and the rate at which this energy is produced depends on the circumstances. In the case of an emergency d.c. powered operating theatre light, shown in Fig. 14.4, the energy production of heat and light could be 96 watts or 96 joules of energy per second.

In order to drive the electric current through the light bulb, a potential difference must be present across the bulb. The unit of

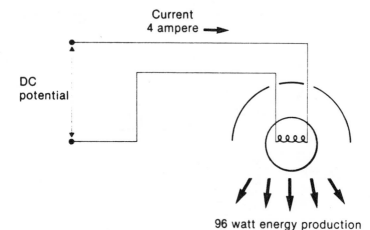

96 watt energy production

Figure 14.4　Example of the definition of potential based on a d.c. powered lamp.

potential difference is the volt, and this is based on the energy production and current flow.

> *The volt* is defined as that potential difference which produces a current of one ampere in a substance when the rate of energy dissipation is one watt.

Run 1 Amp current through resistor - get

In other words:

$$\text{Potential difference in volts (V)} = \frac{\text{Power in watts (W)}}{\text{Current flow in amperes (A)}}$$

In the example shown in Fig. 14.4:

$$\text{Potential difference} = \frac{96 \text{ W}}{4 \text{ A}}$$

$$= 24 \text{ V}$$

An alternative definition of the volt based on electric charge is sometimes used.

HEAT PRODUCTION AND A.C. UNITS

In addition to its use to define potential, heat production from electric current flow is also used as a basis for comparable a.c. units of measurement. On the left of Fig. 14.5 is a graph of the potential difference across an item of electric equipment supplied by alternating current. A graph of the current flow would be similar.

Amplitude is normally used as a measure of the size of a sine wave (Chapter 13), but a problem arises in using this simple method in the case of alternating current. An alternating current with a maximum

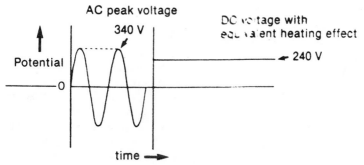

Figure 14.5 Relationship between an a.c. and d.c. supply of equivalent heating effect.

amplitude of one ampere has a smaller heating effect than a constant direct current of one ampere because, in the case of alternating current, the peak flow occurs for only a small fraction of the time. When electricity is used commercially it is the heating and energy production that are of interest. Some units therefore are necessary to relate the current and voltage values of a.c. electricity to the d.c. units so that the heating effects are the same. The units used are root mean square (r.m.s.) values.

An r.m.s. voltage of 240 V (the mains potential in the United Kingdom) has an identical heating effect to that of a d.c. voltage of 240 V when applied across an identical resistor, but its peak voltage is 340 V. The detailed mathematics of r.m.s. values are beyond the scope of this book, but Fig. 14.6 illustrates the meaning of the term.

If all the values of the sine wave are squared, all the amplitudes are converted to positive numbers. By taking the mean of these, a value is achieved which is related to the amplitude of the wave. Finally, by taking the square root of this figure, the equivalent d.c. value is obtained.

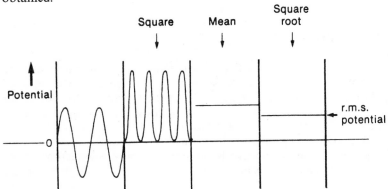

Figure 14.6   Derivation of the root mean square potential.

HEAT PRODUCTION AND FUSES

The rapid increase of heat production with increasing current flow is utilised in the operation of fuses. Essentially a fuse is a wire of a diameter chosen so that if the current exceeds the rating of the fuse the wire rapidly heats and then melts, so stopping the flow of current. Fuses are incorporated into important circuits in electrical equipment so that if a fault develops, in which a current which is higher than normal flows, the fuse element melts and disconnects the circuit. For the safety of electrical apparatus it is essential that the correct size of fuse is used.

ELECTRICAL BURNS

A shock from mains electricity carries a serious risk of ventricular fibrillation (Chapter 16) but sometimes the current passes through only a limb, giving local heating and burns. For example, if a high voltage source is touched, burns could occur at the finger tip where the current flow is concentrated through a small area. The current flow per unit area is known as current density and the greater the current density, the greater the heating effect.

DIATHERMY

The concept of current density is important because it explains the control of the heating effect of diathermy.

The passage of direct current or low frequency alternating current through the body may cause physical sensation, stimulate muscular contraction and give a risk of ventricular fibrillation. These effects become less as the frequency of the current increases, being small above 1 kHz and negligible above 1 MHz, but the heating effect and burning can occur at all frequencies.

Diathermy equipment is used to pass a current of a high frequency, typically 1 MHz, through the body in order to cause cutting and coagulation by local heating of body tissues.

The degree of burning produced depends on the current density. In diathermy equipment there are two connections to the patient, the neutral or patient plate and the active or cutting electrode used by the surgeon (Fig. 14.7). The same current flows through both connections and, since in the cutting electrode the current is channelled through a very small area, local heating and burning occurs. No burning should occur at the patient plate, however, as the current flows through a large area. It is crucial that the large area of contact with the patient is maintained at the patient plate. If for any reason the area of contact is reduced the current density is correspondingly increased and there is a risk of burning should the current density become too high.

If the patient plate becomes completely detached from the patient, or is not attached correctly, the diathermy current may flow to earth through any point at which the patient is toucing an earthed metal object. Thus, in these circumstances it is possible for a burn to be produced where a patient's hand touches the operating table. Figure 14.7 also illustrates the presence in the circuit of an isolating capacitor. This and an alternative isolated type of diathermy circuit increase safety, and both are discussed in more detail in Chapter 16.

Another type of diathermy circuit does not require a patient plate. It is known as bipolar diathermy and the electric current travels down

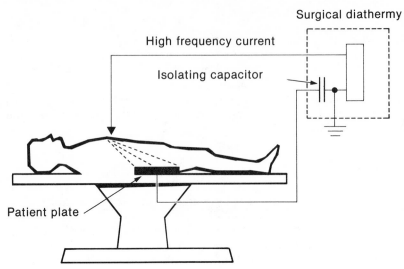

Figure 14.7 Principle of the diathermy apparatus.

one side of a special forceps through the tissues and back through the other side of the forceps. As it is not possible to use high power with bipolar forceps they are used in applications such as neurosurgery in order to localise the current and prevent excessive tissue damage.

PROPERTIES OF ELECTRICAL COMPONENTS

There are standard symbols for electrical components and these are illustrated in Fig. 14.8. The capacitor, inductor and resistor have the properties of capacitance, inductance and resistance, respectively, and these are now considered in more detail.

CAPACITANCE AND ELECTRIC CHARGE

Capacitance is a measure of the ability of an object to hold electric charge, charge being a measure of the amount of electricity. The SI unit of charge is the coulomb (C).

> *The coulomb* is the quantity of electric charge which passes some point when a current of one ampere flows for a period of one second.

Coulombs (C) = Amperes (A) × Seconds (s)

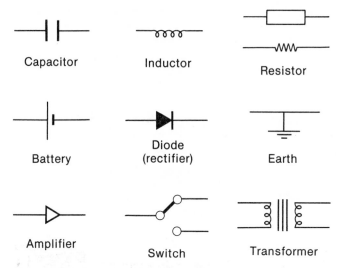

Figure 14.8   **Electrical symbols.**

As the coulomb is an amount of electricity equivalent to 6.24×10^{18} electrons it is obvious that the electron is too small a unit to act as a practical measurement of charge.

The Defibrillator

The defibrillator used for treatment of ventricular fibrillation is an example of an instrument in which electric charge is stored and then released in a controlled fashion. The key component for storing the charge is a capacitor (Fig. 14.9), which consists of two plates separated by an insulator. In Fig. 14.9A the defibrillator is at its maximum setting and a potential of 8000 V is applied across two capacitor plates and this leads to a store of electrons equivalent to 100 mC of charge.

Defibrillators are set according to the amount of energy stored, and this depends on both the charge and the potential. This can be shown as follows:

$$\text{Potential (V)} = \frac{\text{Power (W)}}{\text{Current (A)}}$$

but

$$\text{Power (W)} = \text{Energy (J) per second}$$

and

$$\text{Current (A)} = \text{Charge (C) per second}$$

Therefore:
$$V = \frac{J\,s^{-1}}{C\,s^{-1}}$$

$$= \frac{J}{C}$$

and therefore
$$J = CV$$

In the case of a defibrillator the voltage stored falls as the capacitor discharges and so the energy is only half that calculated from the initial peak potential (V).

$$\text{Stored energy (J)} = \tfrac{1}{2} \times \text{Stored charge (C)} \times \text{Potential (V)}$$

In the example given:

$$\text{Energy} = \tfrac{1}{2} \times 100\,\text{mC} \times 8000\,\text{V} = 400\,\text{J}$$

For treating a patient in ventricular fibrillation, electrodes are applied across the patient's chest and, by means of switches, the 400 joules of energy are released as a current pulse across a patient's chest and heart (Fig. 14.9B). This current pulse, which could be 35A for 3 ms, gives a synchronous contraction of the myocardium after which a refractory period and normal or near normal beats may follow. In practice, other components such as an inductor are included in a defibrillator to ensure

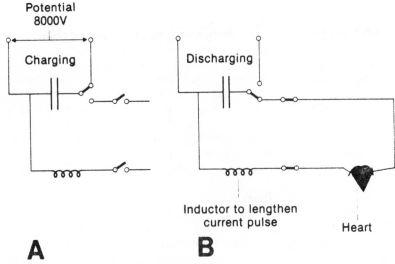

Figure 14.9   Principle of the defibrillator.

that the electrical pulse has optimum characteristics in terms of its shape and duration.

Defibrillators also have lower settings suitable for use with internal cardiac electrodes in a patient with an open chest, and many have provision for synchronised use in treating certain dysrhythmias. When the defibrillator is used in synchronised mode, the energy from the defibrillator must be supplied at the correct time in the cardiac cycle, i.e. during the R wave. If the pulse is mis-timed there is a risk of causing ventricular fibrillation.

Electrostatic Charge

The ability to store a charge is not limited to a capacitor and a charge, usually called an electrostatic charge, can build up on the surface of any object insulated from its surroundings. In the case of a bobbin in a variable orifice flowmeter (Chapter 3), friction may result in electrons being removed from the bobbin surface when it rotates against the wall of the flowmeter. The resulting opposite electric charges on the wall and bobbin can exert a force of attraction leading to the bobbin sticking.

Similarly, insulators may develop charges of this type on their surfaces with risks of sparks, which can be dangerous in the presence of flammable agents (Chapter 23).

CAPACITANCE AND INTERFERENCE

A capacitor cannot directly conduct electrons across the gap between the plates but acts as a total block to direct current. On the other hand, a capacitor is capable of allowing alternating current to pass because this produces changes in charge on the plates so that a small current temporarily flows to and from the plates. In anaesthetic practice, this ability of alternating current electricity to pass across an air gap because of capacitance explains why electrical interference may appear on an ECG trace.

Consider the circumstances shown in Fig. 14.10 where an operating theatre light is separated from a patient by an air gap. Mains electricity at a frequency of 50 Hz is passing through the light. The patient on the table, although not directly connected to the light, acts as one plate of a capacitor with the electric light acting as the other plate. A small mains frequency current passes from the lamp to the patient and unfortunately causes a 50 Hz voltage to appear on the ECG trace. This interference may be of sufficient amplitude to obscure the recording.

Capacitance effects are also used as the working principle of some pressure transducers. In this case, displacement of the transducer diaphragm by pressure causes changes in the capacitor incorporated in the transducer body.

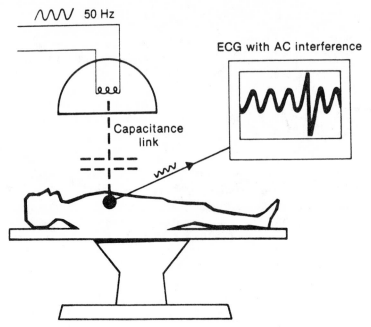

Figure 14.10   Interference on ECG from capacitance.

INDUCTANCE AND INTERFERENCE

Inductance is based on electromagnetic effects and can also lead to interference in biological electrical signals. An inductor is typically a coil of wire as indicated by its symbol (Fig. 14.8). Many components of electrical apparatus have coils of wire with strong magnetic fields around them, e.g. the transformer often found in mains operated apparatus (Fig. 14.11). The magnetic field surrounding such apparatus may cause interference in adjacent electrical apparatus and give another source of interference on the ECG trace or on other biological signal displays. In this illustration the magnetic field formed by the transformer in nearby electrical apparatus induces electric currents in the ECG leads and this source of interference, sometimes called inductive coupling, may lead to interference appearing on the tracing.

Interference from inductive coupling and capacitance effects can be reduced by a system known as screening. The insulated patient monitoring leads are covered by a sheath of woven metal which is earthed, so that interference currents are induced in this metal screen and not in the signal leads. The screening layer is covered by a further layer of insulation.

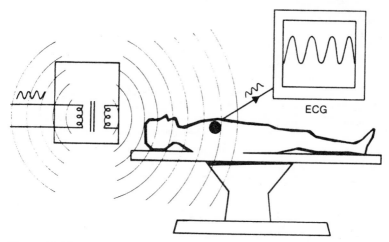

Figure 14.11   Interference on ECG from inductance.

Signal to Noise Ratio

In practice, any electrical waveform or signal includes a certain amount of interference from capacitance and inductance effects and from other sources. This interference is given the term 'noise' from analogy with acoustics, in which the noise may be regarded as unwanted sound of various frequencies.

Whether the noise is a practical problem or not depends on its magnitude in relation to the waveform concerned, a useful measure being the signal to noise ratio. On the left of Fig. 14.12 is an ECG waveform with a high signal to noise ratio, whereas on the right is a similar waveform with a low signal to noise ratio. The signal to noise ratio is improved if the noise is reduced to a minimum by eliminating its

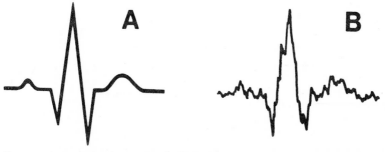

Figure 14.12   Signal to noise ratio. Comparison of high ratio (A) with low ratio (B).

source or by use of differential amplifiers considered in Chapter 15. Unwanted distortion of a biological signal, such as an ECG by diathermy, can be avoided by using an electronic filter to cut out the high frequency diathermy signal.

The signal to noise ratio cannot be improved by amplifying the waveform because both the signal and the noise will be increased by the same factor, but the ratio can be improved by averaging a repetitive waveform, e.g. the signal, over several cycles, and instruments are available to do this electronically. The noise is averaged to zero because of its random nature, whereas the signal remains constant if it is repetitive.

RESISTANCE

The unit of electrical resistance is the ohm (Ω).

The ohm is that resistance which will allow one ampere of current to flow under the influence of a potential of one volt.

$$\text{Resistance } (\Omega) = \frac{\text{Potential (V)}}{\text{Current (A)}}$$

This relationship is known as Ohm's law. Electrical resistance is analogous to flow resistance (Chapter 2), potential taking the place of pressure and current the place of flow. Like flow resistance, electrical resistance varies with the circumstances. A temperature rise increases the resistance of a wire resistor but reduces that of many semi-conductors, this fact being used in temperature measurement (Chapter 9) and in the thermal conductivity detector (Chapter 20).

If a wire is stretched it becomes longer and thinner and therefore its resistance increases. Such resistors are known as strain gauges and are used in pressure transducers as illustrated in Fig. 14.13. In a pressure

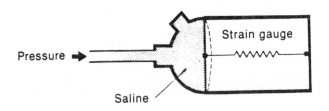

Diaphragm

Figure 14.13   Strain gauge pressure transducer.

transducer of this type, movements of the diaphragm with changes of pressure alter the tension in the resistance wire thus changing its resistance. The changes of current flow through the resistor can then be amplified and displayed as a measure of pressure change.

To monitor or measure such changes of resistance, a circuit known as a Wheatstone bridge is often used. In its original form this consists of a set of four resistors, a source of electric potential—e.g. a battery—and a galvanometer, arranged as shown in Fig. 14.14. In the example, resistor

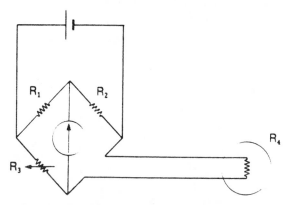

Figure 14.14 Wheatstone bridge circuit, R_4 is a strain gauge transducer or a resistance thermometer and R_3 a variable resistance which is adjusted until there is a null deflection on the galvanometer.

$R4$ could be a strain gauge transducer or a resistance thermometer, while $R3$ is a variable resistance which can be adjusted until the galvanometer reads zero. In this condition the bridge is said to be balanced and:

$$\frac{R_1}{R_2} = \frac{R_3}{R_4}$$

In this form the Wheatstone bridge is a null deflection system, i.e. no current flows through the galvanometer when the bridge is balanced. In practice, however, the output is amplified and connected to a recorder or oscilloscope in place of the galvanometer, and the extent to which the bridge is out of balance is recorded.

Most pressure transducers contain four strain gauges which form the four resistances in the Wheatstone bridge. The transducer is designed so that the resistances of two of the strain gauges at opposite sides of the bridge increase as the pressure increases while the resistances of the other two decrease. This gives a larger potential change at the galvanometer connections of the bridge. The potential is then amplified before it is displayed or recorded.

Figure 14.15 shows another arrangement of resistors R_1 to R_5 with a potential V_1 applied across them. If the five resistors are of equal value, by tapping off from point A to one of the other points B, C, D, or E, a reduced potential may be obtained and in the example potential V_2 is 20% of V_1, i.e. it is attenuated. Sensitivity switches in apparatus often employ a pattern of resistors of this type and in the position shown on the right of the figure potential V_3 is attenuated to 80% of V_1.

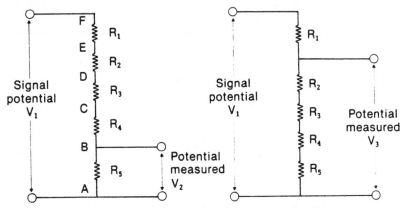

Figure 14.15 Principle of an attenuator.

IMPEDANCE

The resistance of a resistor to the flow of alternating current does not vary with the frequency of the current. A capacitor can transmit alternating current as mentioned earlier, but in this case high frequency current passes through it more easily. With an inductor the reverse is true; low frequency current passes more easily than that of high frequency.

Many circuits and circuit elements behave as though they were made of a combination of resistors, capacitors and inductors and various parts of such circuits possess a particular resistance to the flow of electric current, which in general depends on the frequency of the current flowing. The term impedance is used in preference to resistance when there is a dependence on frequency. The unit of impedance is the same as that of resistance, i.e. the ohm. However, impedance is often indicated by the symbol Z.

A practical example of the use of the variation of impedance with the frequency of the current is the isolating capacitor present in the diathermy circuit. It offers a high impedance to mains frequency current and so protects the patient from electrocution, and is considered in Chapter 16.

Skin Impedance and Attenuation

An important practical example of impedance occurs in the conduction of a biological electrical signal such as the ECG through the patient's skin to the electrode as, in addition to the resistance at the point of contact, capacitance and inductance may affect the signal. Impedance is less if good contact is maintained at the point where the electrode touches the skin. Many different electrode systems are in use but a typical ECG electrode consists of a metal disc covered with conductive electrode gel and fixed to the patient's skin by a surrounding adhesive flange. The electrode gel reduces the electrical impedance at the point of contact with the skin so that this impedance is as small as possible to avoid attenuation of the signal.

The degree of attenuation of the signal depends not only on the skin impedance but also on the input impedance of the monitoring apparatus used to detect it.

Study of Fig. 14.16 and comparison with Fig. 14.15 illustrates the principle. On the left of Fig. 14.16 is an example in which there is a high impedance Z_E from the skin and electrodes. This could arise from dried electrode gel or a loose electrode contact. In addition, there is a relatively low amplifier input impedance Z_A. The signal is greatly attenuated because of the ratio of the two impedances Z_E and Z_A.

The diagram on the right of Fig. 14.16 shows an amplifier with a high

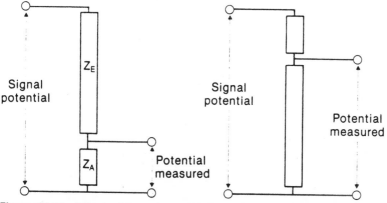

Figure 14.16 Effect of impedance at the amplifier Z_A and at the skin and electrodes Z_E on the measurement of a signal potential.

input impedance used with electrodes with good skin contact to give negligible attenuation of the input signal. Most amplifiers now have a very high input impedance to allow satisfactory ECG recordings even with the simpler and smaller skin electrodes. A high amplifier input impedance also protects the patient against electrocution (Chapter 16).

Skin impedance is lowered when the skin is moist. The degree of moisture depends on sympathetic tone and so skin impedance may be used as a measure of the autonomic activity. This test is known as the galvanic skin response, or GSR.

CHAPTER
15

Biological Electrical Potentials: Their Display and Recording

ORIGIN OF BIOLOGICAL POTENTIALS

The membrane of a cell is composed of phospholipids and has a sandwich structure, with a hydrophobic fatty interior and a protein–carbohydrate exterior. The metabolic mechanisms pump sodium ions out of the cell, and potassium ions in through the cell membrane. These ions carry electric charges, so that the net result is a positive charge of around 90 mV at the exterior compared to the interior of the cell.

As shown on Fig. 15.1, a rapid loss of this potential takes place prior to the contraction of cardiac or skeletal muscle. During this period of depolarisation, sodium ions move into the cell and then potassium ions move out through the membrane. Finally, the membrane recovers its potential as the sodium–potassium pump restores the status quo. Other ion movements also occur. In cardiac muscle, for example, calcium ions as well as sodium enter the cell, so maintaining a longer period of depolarisation than that in the skeletal muscle. Adjacent areas of membrane are destabilised by the region of depolarisation and undergo the same changes so that a wave of depolarisation spreads over the muscle. Similar changes take place at the surface membrane of a neurone to propagate the nerve impulse.

These waves of electrical potential changes are transmitted through the tissues overlying the nerves and muscles, and the signal can be detected by suitable electrodes placed on the skin and displayed as an electroencephalogram (EEG), electrocardiogram (ECG) or electromyogram (EMG). The graphs obtained of voltage against time are complex wave patterns and these waves may be separated into a collection of pure sine waves by Fourier analysis, as described in Chapter 13.

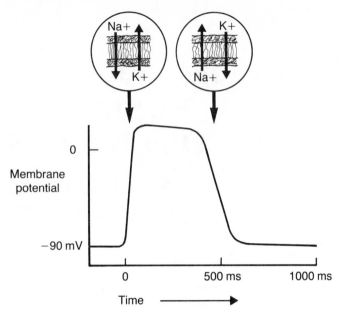

Figure 15.1 Depolarisation changes at cell membrane.

ECG

During a cardiac cycle a wave of depolarisation passes from the atrial pacemaker cells over the atrium and down the AV bundle to spread through the ventricular myocardial syncytium. Potentials from the heart are transmitted through the tissues and can be detected by electrodes to give an ECG recording. These potentials are attenuated as the signal passes through the tissues. Hence the size of the ECG signal detected is only 1 to 2 mV instead of the original potential of about 90 mV mentioned above. The larger the bulk of the cardiac muscle through which the waves of depolarisation pass, the larger the potential detected at the surface, so there is a large QRS complex from the depolarisation wave in the ventricles and a smaller P wave from the atria.

The actual appearance of the ECG depends on the position of the electrodes relative to the heart (Fig. 15.2). Because the atrial signal spreads outwards, the P wave is usually positive regardless of the electrode position. However, in the case of the ventricles the wave of depolarisation travels downwards and to the left, so the QRS complexes vary in appearance according to the electrode position. With an oesophageal electrode positioned close behind the atria of the heart a particularly clear ECG signal of the P waves is obtainable. The time

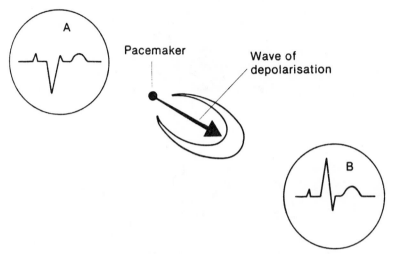

Figure 15.2   The ECG signal picked up by electrodes at A differs from that at B because it depends on the position of the electrodes relative to the wave of depolarisation through the cardiac muscle.

intervals in the complexes are important. For example, a PR interval of over 200 ms indicates delayed conduction from the atria.

Because other muscles give rise to potentials prior to contraction, the patient must relax and make no movement during the recording.

EMG

The electromyogram or EMG records muscle potentials either from surface electrodes or from needle electrodes in the muscle itself. The potentials detected range from about $100 \mu V$ to many mV. Skeletal muscle potentials differ from those of the ECG in that they have a much shorter duration, perhaps 5 to 10 ms, so the EMG gives sharp spikes in place of the complex ECG pattern. The short duration arises because skeletal muscle repolarises more rapidly than cardiac muscle, and also because skeletal muscle has cells which do not polarise in synchrony with adjoining cells. As shown in Fig. 15.3, the amplitude of the EMG signal spike depends upon the number of muscle fibres simultaneously stimulated. This, in turn, depends upon the number of fibres stimulated by the motor neurone. At the top of the figure each motor neurone supplies about eight muscle fibres (for example, in an eye muscle), and so a small spike potential occurs. The example below illustrates a limb muscle with a motor unit of 200 fibres, so a larger spike potential results.

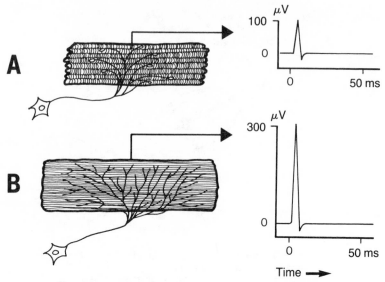

Figure 15.3 EMG potentials from an impulse from the motor neurone supplying an eye muscle (A) and a limb muscle (B).

EEG

In the electroencephalogram, the EEG, 16 electrodes are positioned around the patient's head to pick up the characteristic potentials from the nerve tracts, brain stem and cerebrum. The potentials are much smaller than those of the ECG, being only about 50 μV.

The appearance of the EEG wave pattern is also important. For example, slow, low frequency waves may indicate cerebral hypoxia. In anaesthesia, increasing depth of anaesthesia is usually accompanied by a decrease in the frequency and amplitude of the EEG.

Because the standard EEG is too complex for routine use in anaesthesia, simpler forms of apparatus have been developed. In one of these, the cerebral function monitor, CFM, two electrodes are positioned on the patient's head, and the monitor gives an overall measure of the total cerebral electrical activity within a preset range of frequencies. The CFM does not give a record of the EEG wave patterns. However, a development of it, the cerebral function analysing monitor, or CFAM, includes a display of the activity in the beta, alpha, theta and delta bands.

RANGE OF FREQUENCIES AND POTENTIALS

The relevant ranges of frequencies and potentials of biological signals are shown in Fig. 15.4. In the ECG the frequency range is about 0·5 to 80 Hz while in an EEG the frequency distribution is smaller. The beta waves of the EEG are in the range of 15 to 60 Hz and the slow delta waves start at about 1 Hz. In an EMG there are sharp spikes indicating that an analysis of the wave form would result in a very large frequency distribution of sine waves.

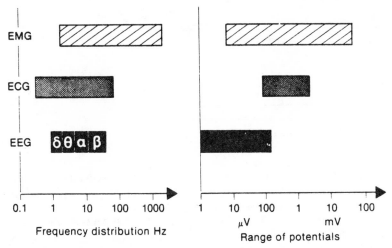

Figure 15.4 Range of frequencies and potentials for biological electrical signals.

THE BLACK BOX CONCEPT

When recording biological variables, the amplifier, recorder and input transducer, if required, may be regarded as a series of black boxes as indicated in Fig. 15.5.

By simplifying the processes of detection, amplification and recording of biological signals in this way, it becomes unnecessary to understand the complicated interactions of individual electrical components in the system. What is essential is to ensure that the electrical characteristics of the signal from the output of any one black box, such as voltage, current and frequency range are compatible with the input of the next black box in the system. Provided that they are matched, the black boxes can be connected together to provide a satisfactory monitoring system.

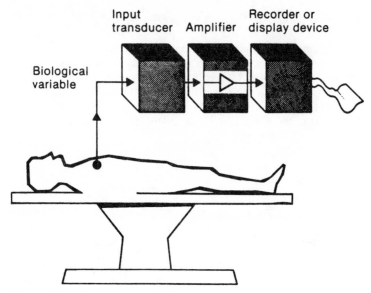

Figure 15.5   Black box concept.

In modern apparatus, components are miniaturised and thus the whole of an amplifier black box becomes a small item incorporated in the recording or display system. An input transducer is needed for arterial or venous pressure measurement or for temperature measurement, but electrodes are used to pick up biological electric potentials directly.

ELECTRODES

Problems can arise at the skin electrodes used to pick up biological electrical potentials (Fig. 15.6). The signal causes a small current to flow through the electrode and this current produces chemical changes at the surface of the electrode, which in turn can alter its impedance and generate a potential, an effect known as polarisation. An extreme example of polarisation arises when a defibrillator is used. The high current flow through the defibrillator electrodes can polarise them severely, and so direct recording of the ECG from these electrodes may temporarily cease after defibrillation.

A second problem at an electrode is that the combination of the metal electrode, the skin surface and its moisture can form a battery, generating a small variable potential. Finally, if the electrode accidentally moves, this may cause the electrical impedance at the skin to alter.

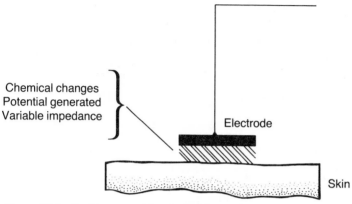

Chemical changes
Potential generated
Variable impedance

Electrode

Skin

Figure 15.6 Problems at skin-electrode junction.

These problems can be solved by the use of a silver electrode in contact with silver chloride on its surface which, in turn is in contact with chloride salt in a gel solution in a spongy pad on the skin. The electrode and pad are securely fixed to the skin with an adhesive disc. The junction between the silver chloride and the chloride ions gives a stable d.c. potential which does not interfere with the recording.

Alternative techniques use electrodes with a large surface area to reduce variability. For example, a sheet of insulating plastic with three separate electrode areas of conducting foil is sometimes used in theatre. It is placed under a patient's back and provides a trace which is adequate for monitoring purposes. In addition, for rapid emergency use, large stainless steel plate electrodes are sometimes clipped to the patient's limbs.

AMPLIFIERS

These are usually differential amplifiers, and measure the difference between the potential from two sources, indicated as S_1 and S_2 in Fig. 15.7. The advantage of this system is that any interference which is common to the input terminals, such as that from 50 Hz mains, is eliminated as only the difference between the terminals is amplified in the differential amplifier. The ability of the differential amplifier to ignore the interference common to both electrodes is known as its common mode rejection ratio. An EEG apparatus, for example, needs a very high common mode rejection ratio as the EEG signal is very small.

In the case of the ECG several leads are used with a switch so that the potential difference between appropriate electrodes may be displayed.

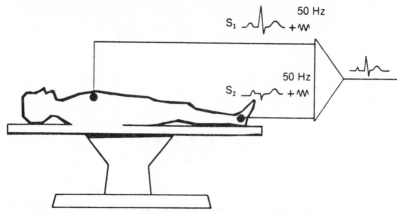

Figure 15.7   Differential amplifier.

For added safety and stability an extra neutral terminal may be used (Chapter 16).

Amplifiers contain some components that are made from semiconductor material. It has already been pointed out that the resistance of semiconductors varies with temperature; hence changes in environmental temperature may cause the d.c. potential at the output of an amplifier to change although the d.c. potential at the input is constant. The problem is known as drift and is much less serious if an amplifier is designed only for a.c. potentials.

The amplifier for a particular biological signal must be able to respond to the range of frequencies concerned (Fig. 15.2). The range of frequencies over which the amplification is relatively constant is known as the bandwidth of the amplifier. Interference in the signal may be avoided if the bandwidth of the equipment used is selected to match the signal correctly. For example, in the EEG, if the upper limit of the amplifier frequency range is restricted, the interference from the high frequency potentials from muscle movement is avoided.

The amplifier must also be suitable for the voltage range of the signal. Thus, a particularly sensitive amplifier is needed for EEG measurement as the size of the signal is so small, perhaps 50 μV compared to the ECG signal of about 1 mV. The ratio of the voltage of the output of an amplifier to the signal voltage at the input is known as the gain of the amplifier. It is measured on a logarithmic scale in units known as the bel. However, the decibel is often used in place of the bel as it is a more convenient unit.

The concept of the bel is also used in sound measurement where it measures the power of the sound present compared to that which would be present at the threshold of hearing.

THE INITIATION OF ELECTRICAL POTENTIALS

In addition to the detection of biological potentials, electrical apparatus is often used to provide an electrical stimulus. One example is the use of the defibrillator considered in Chapter 14. Other examples are pacemakers used to maintain cardiac rhythm, nerve stimulators used in studies of neuromuscular contraction and stimulators used in the treatment of chronic pain conditions. A square wave electrical pulse is used in these cases and the potentials and frequencies depend on the circumstances.

THE CARDIAC PACEMAKER

The stimulating electrode may be inserted into the right ventricle of the heart by cardiac catheterisation or by embedding it in the muscle of the ventricle. The potentials are up to 4 V, and the frequency of the signal is adjusted according to the heart rate required. The duration of the pulse is under 1 ms. Pacemakers and their safety aspects are considered further in Chapter 16.

THE NERVE STIMULATOR

A nerve stimulator produces pulses of potential to stimulate motor or sensory nerves. In the motor nerve stimulator used to assess neuromuscular block, two small pre-gelled silver/silver chloride pad electrodes are applied. The negative electrode, in this case called the active electrode, is placed over the ulnar nerve at the wrist (Fig. 15.8) with the positive electrode a few centimetres proximal. These electrodes

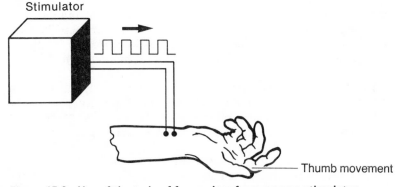

Figure 15.8   Use of the train of four pulses from a nerve stimulator.

give a good electrical contact with a skin-electrode impedance of perhaps 1 kΩ. The stimulator gives a very short pulse of potential of about 200 μs with a peak current of up to 60 mA.

Following each pulse, thumb and finger movement are seen and can be assessed by the strength of the thumb movement measured by palpation or by a strain gauge. The current is adjusted to that needed to give a maximal muscle contraction. A train of four pulses can be given at a frequency of 2 Hz (so that the four pulses cover 2 seconds), and this technique has been found useful to assess the type and degree of nerve block. In partial non-depolarising block, for example, the strength of the thumb movement fades over the four pulses, whereas in complete block, all four twitches in the train are absent. Instead of the train of four technique, tetanic stimulation at a frequency of 50 Hz can be used, but this is painful for a conscious patient.

ELECTRICAL STIMULATORS IN PAIN TREATMENT

Much chronic pain is transmitted by small afferent fibres. One theory suggests that stimulation by larger afferents inhibits transmission or 'closes the gate' on the transmission of pain from the smaller fibres. This effect may take place in the spinal cord and be potentiated by the production of pharmacologically active substances—i.e. morphinones—in the central nervous system in response to stimulation of the larger afferent fibres.

Electrical stimulation of these fibres may be by skin electrodes, nerve stimulation or direct stimulation from electrodes implanted over the dorsal columns of the spinal cord. The stimulator used with electrodes on the skin is sometimes called TNS or TENS (transcutaneous electrical nerve stimulation). The TENS apparatus has small conductive electrode pads to which gel is applied before fixing to the skin.

As in the case of the stimulator used to assess muscle relaxation, the currents used in the TENS stimulator are up to 60 mA, and the pulse durations are also comparable, being between 60 and 380 μs. The pulse frequency is 2 to 200 Hz, depending upon the stimulator and the clinical requirements. When in use the patient feels a sensation of tingling and warmth.

The effectiveness of electrical stimulators varies greatly with different patients. Consequently, a patient must adjust the controls of the stimulator to provide the pulse amplitude, width and repetition frequency that gives him the best pain relief. This should be done under medical supervision as misuse may result in tissue damage.

In addition to the use of nerve stimulators for pain relief, some anaesthetists use them as an aid to identify nerves prior to the administration of local analgesics.

ELECTROCONVULSIVE THERAPY

Another type of stimulator is that used for electroconvulsive therapy, ECT, in the treatment of some psychiatric illnesses. Two saline soaked pads are applied, one to each temple of the anaesthetised patient, and a current of perhaps 850 mA is passed through the brain to induce a major cerebral seizure. The recommended stimulation is in the form of brief pulses of 1·25 ms repeated 26 times a second, i.e. 26 Hz, for a total time of 2 to 5 seconds. The potential will depend on the impedance between the electrodes, but could be about 250 V.

SIGNAL DISPLAY

There are several ways in which biological electrical potentials from the patient can be displayed. If the signal from a patient is steady or slowly changing then an analogue display in which a lever moves over a scale can provide a clear visual indication of the size of a variable. A digital meter works by means of integrated circuits (Chapter 25) and displays values as a set of figures. This can be an advantage when figures are to be noted, and a digital display may also be more accurate and easier to read at a distance. Nevertheless, it is difficult to observe trends on a digital meter. Biological potentials change rapidly, so some form of recording on paper or the use of an oscilloscope screen may be more suitable than a meter in these cases.

The galvanometer, discussed in Chapter 14, is the basis of the usual type of analogue meter which is used frequently but it has shortcomings. There is often no permanent record or tracing and the inertia of the needle and the coil precludes the display of rapidly changing signals. Consequently, meters based on galvanometers are mainly used for displaying slowly changing values, such as a patient's temperature or pulse rate, or average values of other variables.

GALVANOMETER RECORDERS

The galvanometer can be modified to provide a continuous record, one method being to conduct ink through a capillary to a writing point at the end of the galvanometer needle, as shown in Fig. 15.9. An alternative method uses a galvanometer needle with a heated tip so that a record may be produced on heat-sensitive paper, as illustrated in Fig. 15.10.

In these recorders the inertia of the galvanometer needle is increased by the heated stylus tip or by the capillary tube carrying ink. Thus, the speed at which the needle will respond to the signal is limited. These

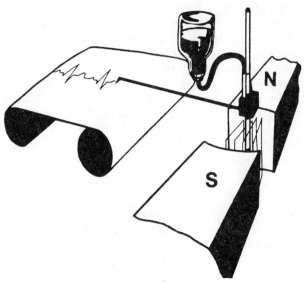

Figure 15.9   Galvanometer recorder.

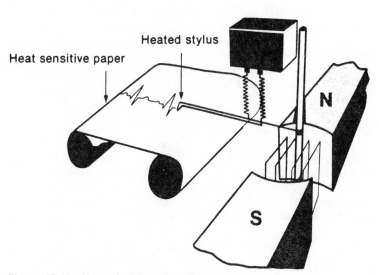

Figure 15.10   Heated stylus recorder.

recorders cannot therefore be used to display a rapidly changing signal such as the EMG. In the case of the heated stylus recorder illustrated, the maximum frequency response is about 80 Hz and is just adequate to cope with the recording of an ECG for clinical purposes.

The total inertia of the moving parts may be reduced if the tracing arm is replaced by a jet of ink as illustrated in Fig. 15.11. The frequency response of such a system increases from the 80 Hz of the heated stylus type of recorder to about 500 Hz. But the disadvantages of such a recorder are the increased cost and the need for greater care in use to avoid blockage of the ink jets.

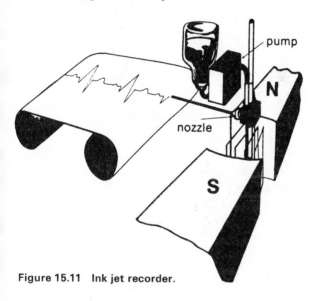

Figure 15.11   Ink jet recorder.

Inertia may be reduced still further if the ink jet nozzle is replaced by a light beam reflected from a mirror, as shown in Fig. 15.12. Ultra-violet light and ultra-violet sensitive paper are used to make the recording. If a permanent record is required, the paper is chemically fixed using fixing solution. The frequency response of such a recorder is in the region of 10 000 Hz, which is adequate to deal with the highest frequency recordings such as the spikes of an EMG. The long light path in this type of recorder also achieves greater sensitivity.

Another problem which can arise in the simpler recorders is related to the fact that the writing point traces an arc (Fig. 15.13). Inaccuracies can arise when measuring the height of the deflection in this case. It is the angle, or arc, through which the pen moves that measures the potential applied to the galvanometer, and the height is not directly proportional to the arc.

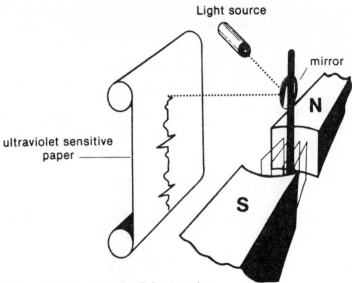

Figure 15.12   Ultraviolet light recorder.

The relationship is indicated on Fig. 15.13. If H is the deflection of the recorder point measured at right angles from the base line, H is r sine A, where r is the length of the recorder pointer and A the angle of deflection. If angle A is measured in radians, arc length S is equal to $r \times A$. Then the error is as follows:

$$Error = H - S$$
$$= r (\sin A - A)$$

Sine distortion would be less if the radius r of the arm of the recorder were increased. However, it is not practical to have a long arm in a simple recorder as this increases inertia.

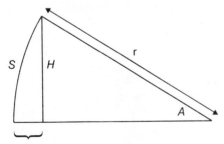

Timing error

Figure 15.13   Sine distortion.

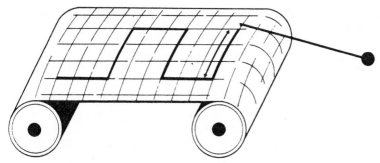

Figure 15.14   Arc distortion of tracing.

In addition to sine distortion there is also a timing error in the direction in which the paper moves (as shown in the figure), and this is related to the cosine of angle A.

With some recorders, the grid on the paper has a curved vertical axis. This helps to compensate visually for the fact that the recorder traces an arc, and solves the problem of the timing error (Fig. 15.14). If the calibration of the grid is appropriately non-linear, this can compensate for sine distortion.

In the case of the heated stylus recorder, this arcing may be corrected by replacing the recorder point with a heated rod system as illustrated in Fig. 15.15.

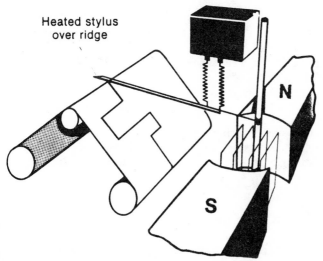

Figure 15.15   Use of a ridge on recording paper to remove arc distortion.

The rod rides over the curved surface of the paper, so that vertical lines will be straight.

In the ink jet recorder the jet plays on the vertical surface of the recording paper, and this removes the curvature in the tracing. Similarly, the ultra-violet light recorder eliminates this problem as the beam is projected to give a vertical tracing.

These systems remove both the grosser curvature and the horizontal timing error which are present with the simpler recorders. Nevertheless, there is still a slight error, known as tangent distortion, when the height of the tracing is used as a measure of the deflection because this height is not directly proportional to the angle of the deflection.

As illustrated in Fig. 15.16, the recorder pointer R in these cases is of

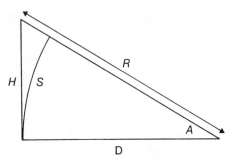

Figure 15.16   Tangent distortion.

variable length as the distance traversed by the ink jet or beam of light varies during recording. The distance of the tracing paper from the pivot of the galvanometer is a constant D, and arc S represents the true deflection of the galvanometer angle A caused by the signal. If A is in radians the error is as follows:

$$\text{Error} = H - S$$
$$= D\,(\tan A - A)$$

As in the case with sine distortion, tangent distortion is less when angle A is smaller, so in recorders such as the one which uses ultra-violet light, with a narrow angle and a long light path, tangent distortion is negligible.

POTENTIOMETRIC RECORDERS

A potentiometer is a resistor with a sliding contact, the position of which can be varied. If connected to a known potential V, a potentiometer may be used to measure an unknown potential, as shown

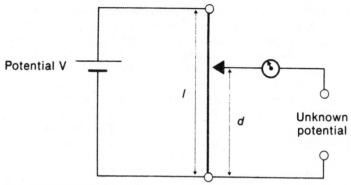

Figure 15.17 Potentiometer.

in Fig. 15.17. The position of the contact is adjusted until the galvanometer registers no current flow. Under this condition, if the distance of the contact is d along a resistance with total length l, then:

$$\text{Unknown voltage} = V \times \frac{d}{l}$$

Figure 15.18 indicates how the potentiometric recorder is based on the principle of the potentiometer. The sliding contact C traces the signal potential but, instead of being adjusted manually, it is moved by

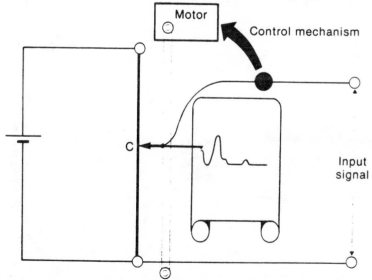

Figure 15.18 Potentiometric recorder.

an electric motor. The motor is driven by an amplifier, which measures the difference between the signal potential and the potential at C. This is done in such a way that the point is always returned to a position on the slide wire where the potential is equal to the input potential; a writing stylus is added to trace the movement of contact C on a chart as shown.

A motor moving a pointer containing ink over a wide chart cannot respond rapidly to sudden changes, so potentiometric recorders have a limited frequency response. They may be used for longer term recordings, such as the patient's temperature, and, in contrast to the narrower tracings usual with galvanometer recorders, a potentiometric recorder can provide a wide tracing with an ink pen system and can also permit overlapping tracings.

If several simultaneous tracings are required, then a number of amplifiers and motors are needed to drive pens with different coloured inks over the chart. The recordings however are offset in time from each other, i.e. their positions on the chart are staggered because space is needed for each moving stylus.

THE OSCILLOSCOPE

The principle of the oscilloscope is illustrated in Fig. 15.19. A hot cathode produces an electron beam which passes through two deflecting devices, one of which deflects the beam on the X axis and one on the Y axis. In the case illustrated these are in the form of simple plates

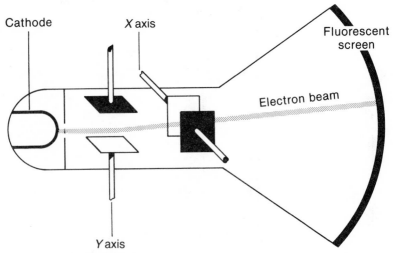

Figure 15.19 Oscilloscope.

which deflect the beam, depending on the potential applied to these plates. The beam then strikes a fluorescent screen where it produces a tracing. A saw-tooth potential deflects the beam in the X direction. The signal, which may be an ECG or perhaps a blood pressure tracing from an arterial pressure transducer, deflects the beam in the Y direction. For an ECG the sensitivity control of the oscilloscope is usually adjusted so that a 1 mV calibration signal gives 1 cm vertical displacement of the trace while the speed control alters the frequency of the saw-tooth potential and so influences the time taken for one horizontal sweep of the trace.

The advantage of the oscilloscope is that the electron beam has negligible inertia, thus leading to a very high frequency response for the apparatus; in addition, there is a continuous display of the signal being monitored. The disadvantage is that there is usually no permanent record with the simpler oscilloscopes.

Memory oscilloscopes are also available. In these, the incoming electrical signal is chopped as it arrives into fragments of a few milliseconds duration, resulting in a series of discrete values. These are stored electronically in a memory bank in digital form. The complete memory store is continually scanned and displayed on the screen. As new fragments of information are added at one end of the memory, old fragments are removed from the other end. At any instant the input of new information to the memory can be stopped, and the contents of the store when read out repeatedly on the screen produce a frozen trace. As a result of the chopping process, the frequency response of a memoryscope can be much less than the conventional oscilloscope , though still fully adequate for ECG monitoring.

The memoryscope uses integrated circuits to process the signal prior to its display and give, in addition, a digital readout on the screen of relevant data. Thus, analysis of the frequency of the QRS complexes of the ECG gives pulse rate, and analysis of the arterial pressure waveform can give systolic, mean, and diastolic pressures. For observation of several tracings and figures simultaneously, colour tracings and figures may be provided.

Modern monitors may also be coupled to computers and recorders to give a written readout of the data. Alternatively, a record can be stored on a magnetic disc for subsequent analysis.

16

Electrical Safety

When the flow of electric current through a person is excessive, there is a risk of electric shock and burns. Electric burns are discussed in Chapter 14. Electric shock and its prevention are considered here.

MAINS SUPPLY, SKIN IMPEDANCE AND EARTH

In order to understand how the risk of electric shock can arise, consider how electricity is supplied to a hospital (Fig. 16.1). A power station supplies electricity at very high voltage to a substation, where the voltage is reduced by a transformer. The current passes between the

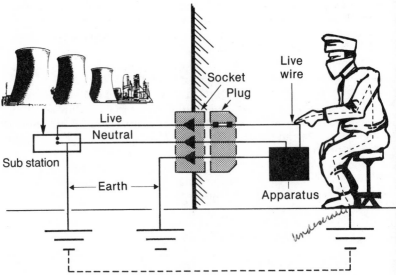

Figure 16.1 Power supply to the hospital and the position of the anaesthetist in completing the circuit between the live wire and earth.

218

substation and the hospital along two wires, the live and the neutral, the neutral wire being connected to earth at the substation. Mains electric sockets in the hospital provide connections to the live and neutral conductors and also to a third conductor, which is connected to earth at the hospital.

If a person touches the live wire in the hospital, an electric circuit can be completed through his body, through the earth, and back to the substation. This could be an anaesthetist touching the live wire in a damaged cable as shown in the figure, so that current passes through him and through his antistatic footwear to the earthed antistatic floor. The effect of the electric shock received by the anaesthetist depends on the size of the current which flows. If this is only a small current of 1 mA, the anaesthetist is likely to feel a tingling sensation on touching the live parts of the apparatus, but he will suffer no serious effects. The mains electricity supply in the United Kingdom is at 240 V potential (in North America and some other countries it is 110 V) and the current which flows through the anaesthetist depends on the impedance he presents to this flow. In the situation described, the impedance of his skin and tissues may be ignored as this is small compared to that of the antistatic footwear and floor. If the impedance of the footwear and the antistatic floor is 240 kΩ, the current is 1 mA, i.e.

$$\text{Current} = \frac{\text{Potential}}{\text{Impedance}}$$

$$= \frac{240}{240\,000} \times 1000 \text{ mA}$$

$$= 1 \text{ mA}$$

Antistatic shoes have a strong protective effect by virtue of their high resistance. It is recommended that the impedance of such shoes should be between 75 kΩ and 10 MΩ when new. This value of impedance is low enough to permit safe dissipation of electrostatic charges, but high enough to give some protection against electric shock.

In Fig. 16.2 the anaesthetist is touching faulty apparatus but is wearing non-standard footwear, and is standing in a pool of saline which he has spilt on the floor and which is in contact with an earthed water pipe. Saline is a good conductor of electricity and so in these circumstances a higher current flows. As the current increases over 1 mA, the sensation of tingling gives way to pain and the current stimulates muscular contraction. In the case illustrated the current could be 24 mA and the anaesthetist would be unable to release the handle of the electrically faulty apparatus. The current therefore, continues to flow, passing through his body to earth through the footwear, the pool of saline and the water pipe. In these circumstances,

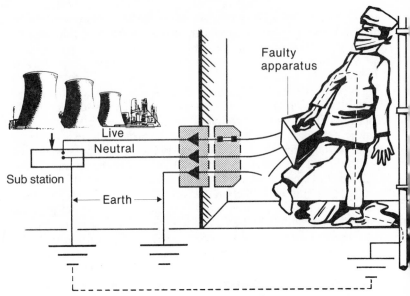

Figure 16.2 Increased risk to an anaesthetist wearing non-standard footwear in a pool of saline.

current need pass through the chest for only a fraction of a second before ectopic beats occur, with the risk of ventricular fibrillation; a risk which increases as the current increases. In this case, the impedance of the footwear is negligible and the impedance of the floor has been bypassed. The internal tissue impedance of the anaesthetist too is small, being only a few hundred ohms. Most of the impedance now occurs at the points of contact of his skin, i.e. his hand with the equipment, and his feet with the shoes. In other words, the skin impedance is the main component, being perhaps a few kilohms. If it is 10 kΩ, the current may be calculated as before, i.e.:

$$\text{Current} = \frac{\text{Potential}}{\text{Impedance}}$$

$$= \frac{240}{10\,000} \times 1000 \text{ mA}$$

$$= 24 \text{ mA}$$

Skin impedance is not constant, however, but depends on many factors. If the anaesthetist's hands are wet, and if the surface area of his hand touching the live equipment is large, the impedance may be much less than 10 kΩ and the current proportionally greater.

The impedances of the contacts with the source of electric current and with the earth are of vital importance when considering the risk of ventricular fibrillation in equipment users and patients. As mentioned above, the skin impedance is not a constant factor. Thus, the impedance is high with dry skin, lower if the skin is damp, and it is lowered further if there are needles or cannulae passing through the skin. Finally, the larger the area of contact with the source of electricity or with the earth the lower is the impedance.

The risk of ventricular fibrillation is increased by additional factors such as the timing of the arrival of an electric impulse at the heart. The risk is much greater if the electric current passes through the heart during the repolarisation of the muscle cells, which occurs during the early T wave of the ECG. Another factor is the form of the electric pulse. Mains alternating current of 50 Hz is more dangerous than high frequency current of 1 kHz or greater. Finally, ventricular fibrillation from electric shock can occur at a lower current in patients with myocardial disease or dysrhythmias.

PROTECTION AGAINST ELECTRIC SHOCK

The risk of electric shock can be greatly reduced if equipment is constructed and maintained to a suitable standard. All medical equipment purchased in the UK, used in the patient's environment and influencing his or her safety should meet the requirements of British Standard 5724: Safety of Medical Electrical Equipment, or International Electrotechnical Commission Standard 601–1.

These standards include the classification of equipment according to the means of protection it provides against electric shock. There are three classes for electromedical equipment.

Class I Equipment

In class I equipment any conducting parts that are accessible to the user, such as the metal case of an instrument, are connected to an earth wire which becomes the third wire connected via the plug to the mains supply socket (Fig. 16.3). If a fault occurs in the equipment in which a connection is made inadvertently between the live supply and the case of the equipment, a circuit is completed from the power source through the case of the instrument and then to earth, so that a high current flows which melts a protecting fuse (or fuses). This disconnects the circuit and removes the live potential from the equipment case.

For this protective system to operate the earth wire must be connected correctly and fuses must be present in the live and neutral wires. In the UK a fuse is also incorporated in the mains plug. The colour code in the cable connecting the mains plug to the equipment is brown for live, blue for neutral, and yellow and green for earth.

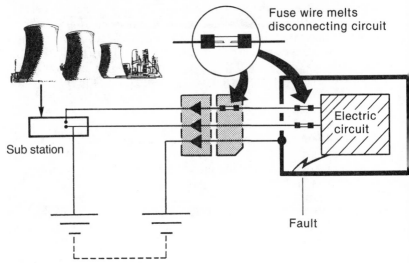

Figure 16.3   Class I equipment. Principle of earthing of the metal case of apparatus and use of a fuse.

Class II Equipment

In class II equipment, also called double-insulated equipment, all accessible parts are protected by two layers of insulation, or by reinforced insulation, so that there is no possibility of a person touching any conducting part that may become live through the occurrence of a fault. An earth wire is not required for class II equipment.

Class III Equipment

In class III equipment there are no potentials exceeding 24 V a.c. or 50 V d.c. Normally, it is not possible for these voltages to produce an electric shock but there may still be a danger of microshock, as discussed later.

SHOCK RISKS WITH EARTHED EQUIPMENT

There may be risks in the theatre of electrocuting the patient if he is allowed to be in contact with earthed equipment of class I type. In Fig. 16.4 the patient's feet have been allowed to touch the stem of a portable lamp, the outside of which is earthed. It provides a direct route through the patient from a faulty live apparatus to earth, the only protection being the size of the patient's skin impedance. If the latter is 10 kΩ, the current through the chest will be 24 mA which is capable of causing ventricular fibrillation.

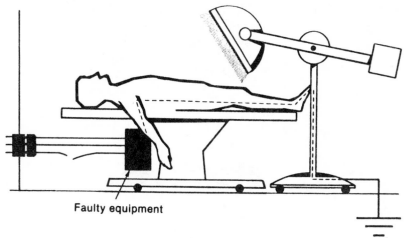

Faulty equipment

Figure 16.4 Risk of electrocution in a patient if he is in contact with earthed apparatus.

Space blankets are made of metal coated plastic and may give an electrical hazard if they come into contact with earthed metal apparatus, thus forming a low resistance path to earth. These blankets are used to reduce hypothermia by reflecting radiant heat back to the patient but they must not be used in theatre, especially if diathermy is in use because of the electrical safety hazard and also because they can burn. In intensive care areas and in proximity to monitoring equipment, they should be used only with care.

Patient electrodes, used for example in ECG apparatus and diathermy, may provide a route to earth and so give an increased risk of electric shock.

DIATHERMY RISKS

In surgical diathermy there are particular risks of inadvertent burns and of electric shock. The principle of diathermy was described in Chapter 14; a high frequency current of about 1 MHz causes burning at the active electrode where current density is high. However, there is no burning at the neutral plate where the current density is low (Fig. 16.5).

Suppose the surgeon activates the diathermy when the active electrode is not in contact with the patient, or when there is poor contact at the neutral plate. Then the normal circuit is broken and current may flow through a different route, such as a point of contact of the patient with the metal operating table, metal equipment or ECG

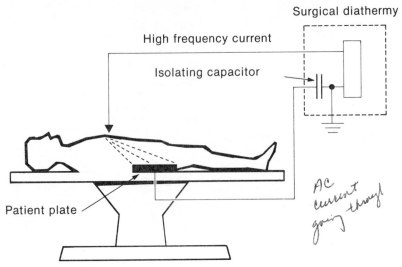

Figure 16.5   Diathermy apparatus—use of isolating capacitor.

electrodes. Burns may then occur at these points of contact. Even in the absence of direct contact, diathermy current can flow by capacitance linkage (Chapter 14). Because diathermy current has a high frequency, capacitance links offer much lower impedance to its flow and such capacitances between the patient and earth or between the patient and the diathermy leads give alternative routes for current flow. A thin layer of insulating drapes may not protect the patient from the risk that diathermy current could flow and cause burns.

The problem described is reduced if the patient plate, and hence the patient, is kept at earth potential. However, a simple earth connection, as used in older machines, increases the risk of electrocution. The isolating capacitor shown in Fig. 16.5 minimises these difficulties. In this case the lead from the patient plate is earthed but the isolating capacitor shown has a high impedance to the low frequency 50 Hz mains current, and so protects the patient from electrocution. Nevertheless, it has a low impedance to the 1 MHz diathermy current, so that all stray currents flow through the neutral plate and do not cause burns. Consequently, the risk of burns at points of accidental contact with earthed metal surfaces is reduced, but this system works only if the neutral plate is in good contact with the patient.

ISOLATED PATIENT CIRCUITS

An alternative system used to reduce the risk of electrocution and burns with diathermy is known as the isolated or floating patient circuit; also known as an earth free circuit (Fig. 16.6). The electric circuit formed by the diathermy output, the cutting or active electrode and the patient and neutral plate have no connection to earth. There is, therefore, little risk of alternative paths for the current through earthed apparatus. Although the equipment is designed and constructed to keep capacitance links very small, stray capacitance linkages cannot be eliminated completely. Hence there is still a risk of burns if the diathermy is operated incorrectly.

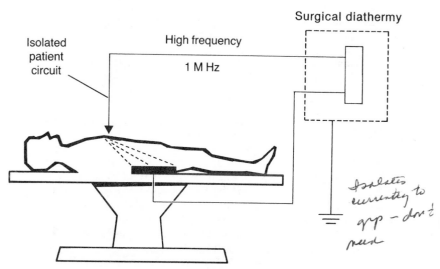

Figure 16.6   Diathermy apparatus—use of isolated circuit.

With either of the two types of surgical diathermy described, risks can be minimised by ensuring good connection of the neutral plate to the patient, by not activating the diathermy until the active electrode is in contact with the patient's tissues, by spacing ECG electrodes and other electrodes away from the area where the diathermy is being used, and by ensuring that the patient does not come into contact with earthed apparatus.

Another point at which the patient used to be earthed was through ECG equipment. For ECG recording an extra lead, known as the neutral lead, is often used to reduce interference in the ECG waveform. In earlier monitoring apparatus this lead was attached to the casing, and earthed as shown in Fig. 16.7A. This had several disadvantages. It increased the risk of electrocution if the patient accidentally came into contact with a source of mains electricity. Burns, too, could occur at the

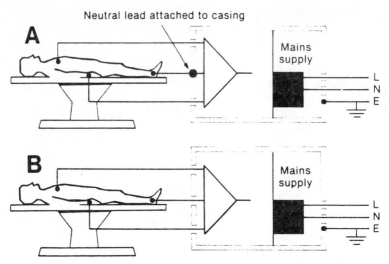

Figure 16.7   Risk of electrical shock in older ECG apparatus (A) is avoided in present earth-free apparatus (B).

ECG electrode if this system was used with diathermy in the presence of a faulty diathermy earth. Modern ECG apparatus avoids these risks and problems by using an isolated or floating patient circuit, as shown in Fig. 16.7B. The isolation is such that, even if mains potential arises between the patient electrode and earth, the equipment will not be damaged and excess current will not flow.

Patient monitoring equipment often uses such isolated circuits to minimise the risk of electric shock to the patient. The amplified signal from an isolated patient circuit can pass into the rest of the equipment through transformers which ensure electrical isolation of the circuits and, as shown in the figure, all leads are insulated from earth.

The concept of the isolated patient circuit is an additional factor in the classification of equipment and can apply to any of the three classes described previously. In class II equipment with a floating circuit, not only is the power supply circuit insulated from all accessible parts but the patient connected circuit itself is isolated and insulated from the rest of the equipment. With such techniques there is little risk of gross mains shock to the patient or operator. Nevertheless, as described in Chapter 14, the alternating current of mains electricity can produce small currents and potentials by inductive and capacitive coupling. These currents can be induced in the metal case of an instrument, or in other circuits or even in a person touching insulated equipment. Such currents, known as leakage currents, will not cause gross electric shock but in some circumstances can give a danger known as microshock.

MICROSHOCK

Ventricular fibrillation resulting from electric shock is caused by the passage of an electric current through the myocardium. If the electric current passes from a hand to the feet (Fig. 16.4), it not only passes through the heart but also through a portion of the trunk. Therefore, the total current flow through the heart itself is only a fraction of the 24 mA which may pass from the hand to the feet. It is the fraction of the current passing through the myocardium or the current density (current per unit area) in the region of the myocardium that determines whether ventricular fibrillation will occur.

If there is a faulty intracardiac catheter passing from an item of monitoring equipment into the heart itself (Fig. 16.8) and if this catheter touches the wall of the heart, any electric current flowing along the catheter will pass through a very small area of the heart. In such a case, a current of only 150 µA may produce the same current density in a portion of the myocardium as that produced by 24 mA flowing from a hand to the feet. Hence, ventricular fibrillation may be induced. This is the phenomenon known as microshock.

As with gross electric shock, the severity of microshock increases as the frequency of the current decreases, so the risk is greatest at low frequencies such as mains frequency and with direct current.

Note that it is not necessary for the faulty apparatus to be at mains potential for there to be a risk of microshock. As microshock currents are at levels of microamperes instead of milliamperes, the potentials can

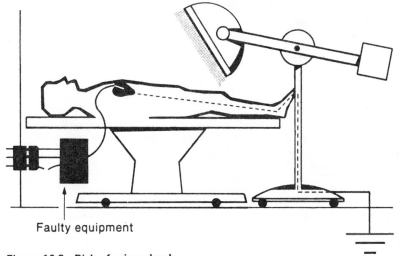

Faulty equipment

Figure 16.8   Risk of microshock.

be proportionately scaled down and there is a risk of microshock from a potential even lower than 1 V. In the case illustrated in Fig. 16.8 equipment attached to the catheter is faulty and a small leakage current is flowing through the catheter, through the heart in the earthed patient. In the illustration the patient is earthed by his feet being in contact with a metal lamp but, alternatively, he might be earthed by contact through the operating table or through faulty equipment. Although the impedance to earth through these items may be high enough to protect against major electric shock, this may not prevent the flow of a small current giving the risk of microshock.

The anaesthetist too may act as an earthing point for the patient if he is in contact simultaneously with the earthed casing of apparatus and with the patient (Fig. 16.9). In this event the small current from the faulty equipment travels via the cardiac catheter to give a risk of microshock in the patient even though the current density in the anaesthetist is too small to be noticed by him.

Microshock is a risk in patients who are fitted with an intracardiac pacemaker with an external lead. A temperature-monitoring probe placed in the lower third of the oesophagus behind the left atrium also gives a possible route for electricity to pass in close proximity to the heart, although it may not be as dangerous as a catheter placed directly in the cavities of the heart.

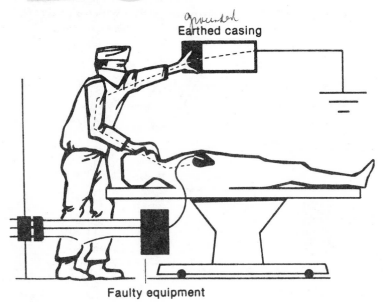

Figure 16.9 Anaesthetist as a potential earthing point for the patient.

LEAKAGE CURRENT STANDARDS

In addition to the classification of equipment into three classes, with or without floating circuits, electromedical equipment is tested and classified according to the size of leakage currents which can arise. Exacting standards have been defined based upon assessments on equipment operated with faults, such as reversal of the neutral and live wires or with disconnection of the earth.

Equipment used with electrodes which may contact the heart directly is termed type CF, indicating that it is for cardiac use and has a floating circuit. In this case, the leakage current through its intracardiac connection must be under 50 µA, even if it is operating with a single fault.

Other medical monitoring equipment is termed type B, or type BF if it has a floating circuit. For these types of equipment, the maximum permitted patient leakage current is higher than for CF equipment, being 500 µA under single fault conditions.

The value of the maximum permitted leakage currents varies over the range of conditions in which they can be measured and the two values quoted are only examples.

Whenever new equipment is received into a hospital it should be subjected to an acceptance test which will verify that leakage currents and other characteristics affecting electrical safety are within the allowed specifications. Equipment should also be serviced regularly to ensure that its electrical safety does not become impaired.

CARDIAC PACEMAKERS

Special risks apply to patients with cardiac pacemakers. Two types of pacemaker are used. Firstly, for temporary use over a few days, a special pacemaker catheter can be inserted through the subclavian or neck veins into the right ventricle. X-ray control is desirable, and full resuscitation facilities must be available during the procedure. Pulses of potential are then applied through the electrode, the potential being adjusted to the minimum needed, usually under 4 V, the pulse duration being under 1 ms.

The second form of pacemaker is for longer term use. The battery powered pacemaker and its pacing lead are embedded in the patient's tissues. Instead of an endocardial pacing lead, electrodes may be embedded in the myocardium, but then cardiac surgery is needed with transmediastinal exposure of the heart. Modern pacemakers are normally designed to work in a demand mode. This means that the stimulating electrode has a second function; it senses the electrical complexes from the atria and ventricles. The pacemaker then paces the heart only if the patient's heart rate is outside an acceptable range.

A problem with demand mode pacemakers is that electromagnetic fields from sources outside the patient can cause interference (Fig. 16.10). The electrode of the pacemaker can act as an aerial and pick up signals from outside, then misinterpret them as QRS complexes. There are numerous possible sources of such interference; for example, electric motors, microwave ovens, and the antitheft devices found in shops and libraries. In the operating theatre, diathermy, nerve stimulators and some monitoring equipment may carry risks. Some physiotherapy equipment can also interfere with the correct function of the pacemaker.

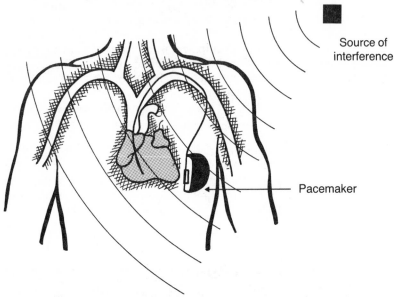

Figure 16.10 Interference at a demand mode pacemaker from an external electromagnetic field.

Patients may experience dizziness and syncope when interference of this sort is present. Even the potentials from adjacent chest muscles can be picked up by the pacemaker and inhibit it. Many demand mode pacemakers have an inbuilt safety factor in that they change to a fixed rate mode when excessive interference is present. In addition, bipolar electrodes are available which have a connection to the atrium and ventricle and give more reliable detection of the ECG signal.

CONDUCTING FLUIDS

Most of the infusion and other solutions used in hospitals are good conductors of electricity. These solutions are often used with or near to medical electrical equipment (e.g. infusion pumps), and great care should be taken to ensure that this is not splashed and that liquid is not allowed to enter inside it. If solutions enter electrical equipment they may cause serious malfunction, and can be hydrolysed into an explosive mixture of hydrogen and oxygen by the passage of an electric current.

POWER FAILURE

Failure of the mains electricity supply is another risk to patients, especially those on monitoring and life support systems in theatre and intensive care areas. Most hospitals have provision for automatic connection of power supplies to an emergency generator, should mains failure occur. This risk is minimised further with some monitoring equipment which is battery operated, and ventilators which are powered from gas cylinders or a piped supply.

CHAPTER 17

Blood Pressure Measurement

PRESSURE AT THE HEART

Blood pressure arises from the force of contraction of the myocardium acting on the blood contained in the heart. The simple concept of pressure as force per unit area is less clear in this case because the force of contraction of the muscle is acting tangentially at the heart surface. This force gives rise to tension, tension being (in this context) force per unit length and Laplace's law (Chapter 2) applies.

Figure 17.1 shows a simple model of a spherical ventricle of a heart with a radius R. The pressure P resulting from tension T in the ventricular wall is determined by Laplace's law for a sphere as follows:

$$P = \frac{2T}{R}$$

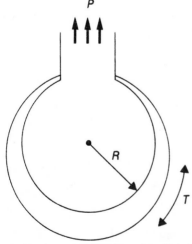

Figure 17.1   Application of Laplace's law to the heart.

The practical consequence of Laplace's law is that as a distended failing heart has a larger radius than normal the distended myocardium is at a disadvantage in that the pressure produced falls unless the muscle contracts proportionately more forcefully. In the normal heart, however, slight distension caused by an increased venous return and the consequent lengthening of cardiac muscle increases the force of contraction (Starling's law of the heart) and the pressure is maintained.

The mechanical work of the heart for a single beat is related to the product of the pressure, and the stroke volume (Chapter 8). Consequently, an increase of blood pressure or of stroke volume increases the work of the heart. Just as the elastic tissue in the lung stores the work of inspiration to power expiration, so the elastic tissue in the aorta and main vessels stores the work of the heart in systole to maintain blood flow during diastole.

PRESSURE IN THE CIRCULATION

Blood pressure is not constant but varies with the site of measurement. Because the circulation is a system of tubes filled with a liquid, hydrostatic effects are present (Fig. 17.2).

In a tall man who is standing, the mean blood pressure could vary from 53 mmHg (7 kPa) at his head to 202 mmHg (27 kPa) at his feet, compared with a mean pressure of 90 mmHg (12 kPa) at his heart. This can be appreciated from Fig. 17.2, if it is remembered that 7·5 mmHg

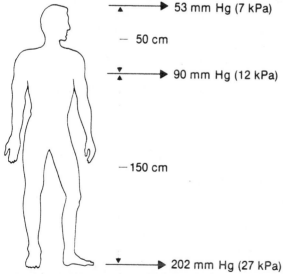

Figure 17.2   Hydrostatic effects upon mean blood pressure.

(1 kPa) equals $10\,cmH_2O$ pressure. In addition to these hydrostatic effects, the blood pressure also depends on the calibre of the vessel and its distance from the heart. Because there is a fall in pressure as a fluid flows through a tube, it might be expected that the mean pressure in arterioles would be lower than in arteries. This simple concept is complicated, however, by the fact that blood pressure is pulsatile and changes in the wave pattern in different vessels alter the systolic pressures in them, as described later. In addition to changes in pressure depending on posture and site of measurement, blood pressure also shows a diurnal variation, being lower during sleep. Minor changes of pressure occur too during the respiratory cycle and more marked changes during intermittent positive pressure ventilation, when, in addition to cyclical changes, there may be a general reduction in blood pressure due to the effects on venous return. Anxiety raises the blood pressure, so readings on three separate occasions may be required to obtain a more representative value.

The total flow in the circulation, or the cardiac output, depends on the peripheral resistance and if laminar flow is assumed:

$$\text{Peripheral resistance} = \frac{\text{Mean (aortic) blood pressure}}{\text{Cardiac output}}$$

Therefore:

$$\text{Mean blood pressure} = \frac{\text{Peripheral resistance}}{} \times \text{Cardiac output}$$

Thus, in a patient with a normal or low peripheral resistance (e.g. with a warm, pink skin) a high blood pressure indicates that a high cardiac output is likely, whereas a high blood pressure in a patient with a high peripheral resistance (e.g. with a cold, pale skin) may not indicate a raised cardiac output. In the formula the mean blood pressure is used. To obtain a fully accurate mean pressure an integration technique would need to be applied to an arterial pressure trace. However, in practice, mean pressure can be estimated as diastolic pressure plus one-third of pulse pressure.

MEASUREMENT OF BLOOD PRESSURE

The simplest non-invasive system of blood pressure measurement consists of an inflatable cuff connected to a manometer. The cuff is placed on the upper arm and the pressure at which the pulse returns after occlusion is identified by a detector system (Fig. 17.3).

Cuff
The cuff must be positioned so that the centre of its bladder is on the medial side of the arm over the brachial artery. The width of the cuff

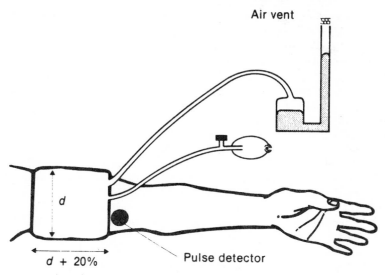

Figure 17.3 Non-invasive measurement of blood pressure.

should be 20% greater than the diameter of the arm d. The cuff with its tubing and connections should not leak.

Manometer

The mercury type of manometer must read zero correctly before use, and it should be used vertically unless it is of the type calibrated to be used at an angle. Partial blockage of the air vent or of the connecting tubings may lead to inaccurate or sluggish readings. Although aneroid gauges are more portable, their calibration should be checked regularly.

Detector

The cuff is inflated to a pressure above the expected systolic pressure, and this is then slowly released at a rate of 2 to 3 mmHg per second. The systolic pressure is indicated by the reappearance of the peripheral pulse, which can be monitored by a detector, the simplest form of which is manual palpation of the radial pulse. Auscultation over the brachial artery at the elbow allows detection of this point by the first-phase Korotkoff sound, i.e. the point at which sounds from blood flow in the artery first appear. The second phase is a slight muffling and the third a rise in volume of the auscultation sounds. The fourth phase of Korotkoff sound is an abrupt fall in sound level and is sometimes taken as representing diastolic pressure, although the final loss of all sound (fifth phase of Korotkoff) is more widely accepted as indicating diastolic pressure.

A microphone with appropriate electronic circuitry is sometimes used in place of a stethoscope, particularly in apparatus intended for use by non-medical or paramedical staff.

In babies it may be difficult to hear the Korotkoff sounds and so ultrasonic or other specialised techniques described later may be needed. A flush technique has also been used in infants but may be less accurate. In this the arm or foot is exsanguinated by means of an Esmarch bandage and the cuff is inflated above the expected systolic pressure. After removing the Esmarch bandage the cuff is slowly deflated until a flush appears indicating systolic pressure.

In anaesthesia a pulse detector is often used on a patient's finger, and may be used in conjunction with the blood pressure cuff. A popular type of pulse detector uses a photocell to pick up transmitted light from a bulb shining through the patient's finger. Pulsatile blood flow in the capillaries between systole and diastole modifies the intensity of light transmitted. Another system uses light reflected from the patient's finger.

Three specialised detector systems—the oscillotonometer, the automatic single cuff system, and the ultrasonic blood pressure apparatus—warrant specific consideration.

THE VON RECKLINGHAUSEN OSCILLOTONOMETER

In this apparatus (Fig. 17.4), in addition to the usual blood pressure cuff, there is a second pulse-detecting cuff with its own sensitive aneroid

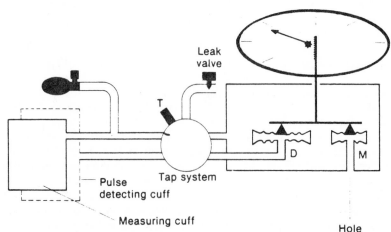

Figure 17.4   The Von Recklinghausen oscillotonometer.

capsule D. This additional cuff detects the reappearance of the pulse in the arm after release of the occluding pressure.

Bellows M in Fig. 17.4 measure the pressure in the measuring cuff in the standard way. It will be seen from the figure that the interior of this particular aneroid bellows communicates to the outside of the oscillotonometer through a channel and hole, while the cuff pressure is transmitted to the interior of the oscillotonometer case and acts on the surface of the bellows M distorting it to allow pressure measurement through a lever system.

The pulse-detecting cuff is distal to the measuring cuff and communicates with its own pulse-detecting bellows D. After inflating the cuffs to above the expected systolic pressure, tap T is adjusted to bring the bellows D into operation, and to permit a slow leak from the system. When systolic pressure is reached, these bellows cause oscillation of the needle on the dial and release of tap T centres the needle to the systolic pressure reading. If tap T is again adjusted to the pulse-detecting position, oscillations continue until diastolic pressure is reached, when they become weaker. The rate of the leak in tap T is adjusted by the leak valve (see Fig. 17.4) to about 2 to 3 mmHg per second. An important point to note when using an oscillotonometer is that the cuff tubings must be correctly connected, and the cuff must be positioned so that the tubings issue from the lower margin. The oscillotonometer should be checked and serviced regularly to ensure the calibration is correct and that there are no leaks. The point at which the oscillations start with this instrument is sometimes difficult to define accurately, and its accuracy has been challenged. Nevertheless, it allows repeated measurements at short intervals and is convenient when access to the patient's arm is restricted.

AUTOMATIC SINGLE CUFF SYSTEM

This system uses a single cuff for arterial occlusion and pulse detection, inflation and deflation of the cuff being controlled automatically. Movement of the arterial wall is transmitted to the cuff, causing pressure changes which are detected by a transducer in the processing unit to which the cuff is connected. Above systolic pressure and below diastolic pressure, oscillations of pressure produced in the cuff by movement of the arterial wall are negligible. As the cuff is deflated, however, these oscillations begin when systolic pressure is reached. They increase until mean arterial pressure is reached and then decrease until diastolic pressure occurs. The processing unit is thus able to display systolic, mean and diastolic blood pressure. This system gives reliable and accurate readings in adults and in children provided the correct cuff size is used.

ULTRASONIC BLOOD PRESSURE APPARATUS

In Chapter 13 the principle of the Doppler effect is discussed, and its use to measure blood flow is mentioned. The probe detects the Doppler change in frequency in the ultrasound waves reflected from moving red cells. This type of probe can be applied with the appropriate gel over an artery, and so used as a pulse detector to facilitate blood pressure measurement, especially in infants.

An alternative instrument, the Ultrasonde, is based upon the Doppler shift in frequency in ultrasound reflected from the arterial wall and is also suitable for use in infants. The system is illustrated in Fig. 17.5. High frequency sound is beamed at an artery, such as the brachial artery. When the artery is either fully open or fully closed the sound waves are reflected off with very little change in frequency. Consider now what happens when blood pressure is being measured. As the pressure in the occluding cuff is released, the arterial wall pulsates strongly between systole and diastole, and these movements of the arterial wall cause Doppler shifts in the frequency of the ultrasound waves reflected. The apparatus is designed so that these changes give rise to an audible noise. The systolic pressure is thus indicated by the point at which this noise begins, and the diastolic pressure when it stops.

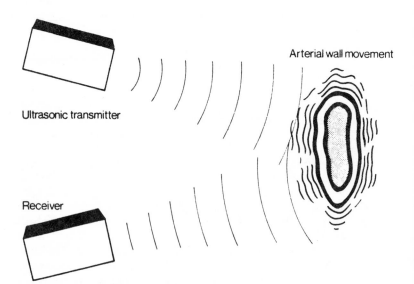

Figure 17.5   Principle of the Ultrasonde.

AUTOMATED MEASUREMENT

Both ultrasound apparatus and the single cuff system can be auto-mated. However, automation does not obviate the need for regular servicing and calibration checks.

In these automated instruments a pump inflates the cuff to a suitably high value, e.g. 160 mmHg, or, in repeat readings to 25 mmHg above the previous systolic measurement. During a reading, the pressure is slowly released at 2 to 3 mmHg per second until diastole is reached and then the cuff is completely deflated. Pulse detection may be based upon the pressure oscillations in the cuff or the ultrasound system and microprocessors are used to enable the equipment to analyse the signal and identify the systolic, mean and diastolic pressures and the pulse rate. The apparatus may also have alarms to indicate if the pressure is above or below preset limits. Microprocessors also allow the apparatus to identify extraneous noise caused by artefacts such as arm movement, and to retake erroneous readings. Such movements should, however, be avoided if possible, as they delay recording.

Repeated readings are possible at regular intervals, with automated apparatus, but reinflation of the cuff at short intervals can give false readings due to congestion of the arm and are uncomfortable to the conscious patient. Usually, readings at 2·5 or 5 minute intervals suffice for most anaesthetic practice. For specialised anaesthesia, however, continuous recording by invasive techniques may be needed.

INVASIVE TECHNIQUES

If an artery is cannulated a direct measurement of blood pressure can be obtained with the help of an infusion system, transducer and recorder.

Ideally, for arterial cannulation, a peripheral artery should be chosen so that the whole limb is not threatened if a clot or haematoma forms. Although some physicians use the brachial artery, the radial artery is usually the first choice but, before cannulation, a modified Allen test is carried out. The patient's hand is clenched into a fist and the doctor occludes both the radial and ulnar arteries with his fingers. Then the patient relaxes his clenched fist and the doctor releases the pressure on the ulnar artery. The patient's hand should then flush within 5 seconds. If flushing does not occur or is delayed, then this indicates that there are poor blood vessel collaterals between the radial and ulnar arteries and, therefore, another artery should be used.

If the ulnar artery is to be cannulated the test is performed similarly, but pressure is released over the radial artery first.

If there are adequate connections between the two arteries, a plastic cannula, e.g. of Teflon, may be inserted into the radial artery. A local anaesthetic is required in the conscious patient and the percutaneous route is usually satisfactory, open exposure rarely being necessary. After insertion, the cannula and its connections must be fixed securely so that there is no possibility of a leak occurring with the risk of severe or even fatal blood loss. To prevent clotting in the cannula, intermittent flushing with heparinised saline through a three-way tap may be satisfactory. However, the high pressures generated by small syringes (Chapter 1, Fig. 1.1) can damage arterial walls or the diaphragm of the pressure transducer. Care should be taken when catheters are being flushed, and syringes smaller than 5 ml should not be used. For longer term recording a continuous flushing system (e.g. 'Intraflo'), as shown in Fig. 17.6, is more satisfactory.

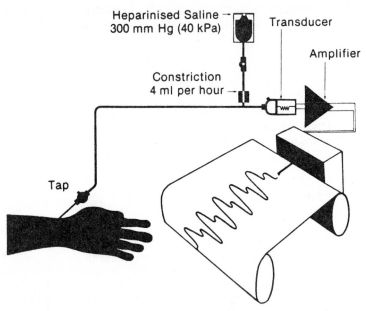

Figure 17.6 Continuous flushing system in arterial pressure recording.

The heparinised saline used is kept in a pressurised container at a pressure of 300 mmHg (40 kPa). It then passes through a drip chamber to a constriction, adjusted so that the flow cannot exceed about 4 ml per hour. This flow continuously flushes the tubings and the arterial cannula. The diagram also indicates the pressure transducer with its amplifier and recorder, and such transducers are considered in Chapter 14.

THE ARTERIAL PRESSURE RECORDING

The final waveform produced may be displayed on an oscilloscope or a recorder tracing. Figure 17.7 shows typical arterial pressure recordings. It can be seen from this that the form of the pressure wave alters as blood flows to the periphery. The blood pressure wave becomes narrower and increases in amplitude in peripheral arteries so that, even with the patient supine, the systolic pressure in the dorsalis pedis artery is higher than in the radial which, in turn, is higher than that in the aorta. This modification of the wave pattern is caused by the change in diameter of the vessels and their elasticity, and possibly also because of the reflection of the wave pattern from the vessel walls.

The systolic and diastolic pressures are easily identified on a tracing and the systolic pressure is found to be an average of 5 mmHg higher with direct measurement at the radial artery than with indirect techniques, while the diastolic pressure is about 8 mmHg lower. The recording also shows the dicrotic notch caused by intra-aortic vibrations.

The frequency range of various biological signals is considered in Chapters 13 and 15. In the case of the arterial pressure wave, the frequency range is 0 Hz to about 40 Hz. The apparatus used for arterial pressure measurement must be able to respond adequately to this range of frequencies. Usually, the amplifiers and recorders have no difficulty in dealing with this frequency range, but problems may arise in relation to the transducer itself and its connections to the cannula.

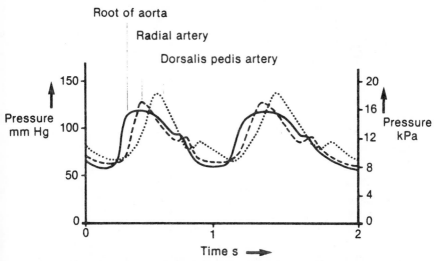

Figure 17.7   Arterial pressure tracings from the aorta, and the radial and the dorsalis pedis arteries.

RESONANCE AND DAMPING

Movement of the diaphragm of the pressure transducer converts the blood pressure change into an electrical signal. This movement of the diaphragm is associated with a very small movement of saline to and fro along the catheter with changes of pressure. Just as a weight on the end of a spring will oscillate at a particular frequency (known as the resonant frequency), so the pressure measuring system consisting of the transducer diaphragm, catheter and saline column possesses a resonant frequency at which oscillations can occur. If this is less than 40 Hz, it falls within the range of frequencies present in the blood pressure waveform. Oscillations occurring at the resonant frequency produce a sine wave which is superimposed on the blood pressure waveform, giving distortion from resonance as shown on Fig. 17.8.

This problem can be avoided if the resonant frequency is outside the range of frequencies present in the blood pressure waveform. The resonant frequency of the combination of catheter and transducer can be raised most easily by using a shorter stiffer catheter, and so the problem occurs chiefly when a long catheter is used.

If there is any restriction to the transmission of the blood pressure from the artery to the transducer diaphragm, the displayed blood

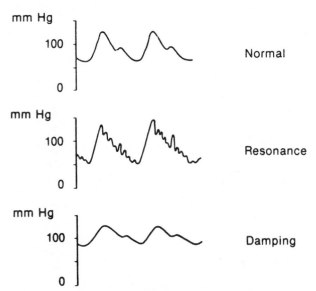

Figure 17.8 Effect of resonance and damping on the arterial pressure trace.

pressure waveform will be damped or smoothed out so that sharp changes are not displayed (Fig. 17.8). Damping of this sort can be produced by air bubbles in the catheter or in the transducer chamber which absorb the pressure change in the saline column. It is also caused by clot formation in the cannula which restricts the movement of the saline column. Both these effects reduce the deflection of the transducer diaphragm, and hence the size of the measured waveform.

COMPARISON OF INVASIVE AND NON-INVASIVE TECHNIQUES

The main advantage of the direct invasive technique of arterial pressure measurement is its potential accuracy. It not only gives greater accuracy by indicating the exact form of the blood pressure trace, but it can also give reliable pressure readings even in the most hypotensive or shocked patient, whereas non-invasive indirect techniques usually fail to record blood pressure below a certain minimum. The invasive method gives a continuous record of the pressure in contrast to the intermittent record provided by non-invasive systems. Hence, it also gives better reliability than the non-invasive techniques if the pressure is continuously varying as, for example, in patients with an irregular or fluctuating pulse rate.

There are, however, disadvantages to the direct technique. There is some risk of arterial damage, whereas the non-invasive methods are harmless to the patient. Direct systems are also more costly than the simpler indirect systems and there is a need for technical skill when using direct methods, while indirect techniques may be carried out by paramedical and even non-medical staff.

CENTRAL VENOUS PRESSURE

The term blood pressure is normally used to refer to arterial blood pressure. However, the venous pressure is also of importance to anaesthetists. Accurate non-invasive measurement of this is impossible, although distended jugular veins in the absence of thoracic inlet obstruction indicate a raised venous pressure.

To measure central venous pressure a catheter is inserted via an arm or neck vein and advanced so that it reaches the superior vena cava. Attached to this catheter are a saline drip, T-piece and a manometer as shown in Fig, 17.9. The techniques used are beyond the scope of this book, but it should be noted that faulty techniques can increase the risk of pneumothorax, haemothorax or hydrothorax and heighten the chance of infection.

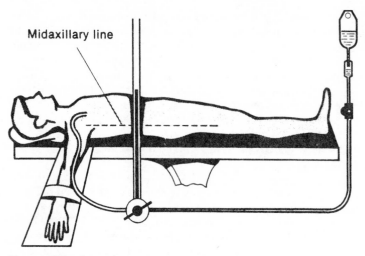

Figure 17.9   Central venous pressure measurement.

Before a central venous pressure reading is taken from the manometer the patient is placed in a horizontal position. Using a spirit level on a rod attached to the manometer, the zero of the scale is set to the level of the midaxillary line, this being taken as representative of the level of the right atrium. The saline drip should be turned off when readings are made and slight movements of the saline level with respiration indicate correct positioning of the catheter. Readings are usually recorded as cmH_2O pressure, but conversion to SI units is simple as $1\,cmH_2O$ is about 100 Pa.

As an alternative to a simple saline manometer, a pressure transducer may be used and this also gives a continuous recording and allows identification of the central venous pressure waveform.

VENOUS PRESSURE AND CARDIAC OUTPUT

The filling of the heart depends on adequate venous pressure. If venous pressure rises, the normal heart fills to a greater extent and automatically increases its output even in the absence of autonomic innervation. The action is described by Starling's law of the heart, and is explained at the beginning of this chapter.

After appreciable blood loss, venous pressure falls and may be used as a guide to transfusion, but care is needed in interpreting venous pressure values as they can be modified by venous tone.

As shown in Fig. 17.10 both the right heart and pulmonary blood vessels are interposed between the central venous pressure recording point and the left heart. Consequently, in a patient with left heart disease excessive intravenous transfusion can give pulmonary oedema before the warning sign of a high central venous pressure is seen. On the other hand, in a patient with lung disease a high central venous pressure may be caused by failure of the right heart. Ideally, both left atrial and central venous pressures should therefore be monitored.

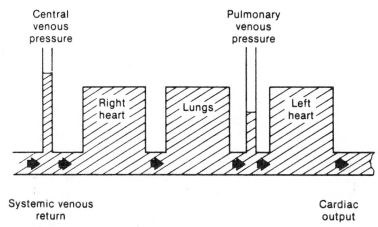

Figure 17.10   Central and pulmonary venous pressures.

Measurement of pulmonary wedge pressure gives an indication of left atrial pressure. The pulmonary wedge pressure is obtained by advancing a special venous catheter through the right heart and pulmonary artery until it wedges into a small branch of the pulmonary artery, where it may be used to measure the pressure in the pulmonary capillaries. The special catheter used has a small balloon at its tip which can be inflated temporarily to facilitate its carriage into the pulmonary artery by the blood stream. It is also inflated during wedge pressure measurement. Continuous pressure recording helps to identify the position of the catheter and ensure that it is positioned in the pulmonary vessels before making the measurement.

CHAPTER 18

Oxygen Measurement

CLINICAL ASSESSMENT

Special techniques are required to measure the amount of oxygen in a patient's blood and in his respiratory gases, because clinical observation of the patient gives only limited information about his oxygenation. For example, although hypoxia can stimulate tachypnoea via the carotid body receptors, this reflex is very variable and can be depressed by anaesthetics and other factors. Furthermore, there are many other possible causes of tachypnoea and it is thus an unreliable indication of hypoxia.

Another indirect indication of hypoxia in the patient is cyanosis. This too has limitations, as different observers may give different assessments of cyanosis and ambient lighting can bias judgment. It is particularly dangerous to assume that the absence of cyanosis indicates adequate oxygenation, because an anaemic patient may have a critically low oxygen saturation and yet not have sufficient deoxyhaemoglobin in the arterial blood to reveal cyanosis. (The term deoxyhaemoglobin is used in this chapter to indicate haemoglobin not combined with oxygen.)

Cyanosis indicates the presence of abnormally high concentrations of deoxyhaemoglobin in the blood in the capillaries, e.g. 15 g litre^{-1} or more. Cyanosis of the lips can normally be assumed to indicate reduced oxygen saturation in the arterial blood. However, in patients with a reduced cardiac output or in capillaries where stasis is present, cyanosis may be unrelated to arterial saturation.

These problems in the clinical assessment of cyanosis are avoided if a sample of arterial blood is taken anaerobically, heparinised and examined in suitable apparatus. If an even more complete assessment of hypoxia in the patient is needed then, in addition, a mixed venous sample may be taken from a central venous catheter.

SPECTROPHOTOMETRIC MEASUREMENTS OF OXYGEN SATURATION

A spectrophotometric technique may be used to assess the amount of deoxyhaemoglobin in blood, the instrument used being known as an oximeter. In the oximeter, light transmitted through the haemolysed blood sample is analysed by means of filters and photocells to obtain the oxygen saturation.

Figure 18.1 illustrates the physical principle underlying the working of the instrument. The light absorbed by blood depends on two main factors, the ratio between the deoxyhaemoglobin and oxyhaemoglobin, and the wavelength of the light.

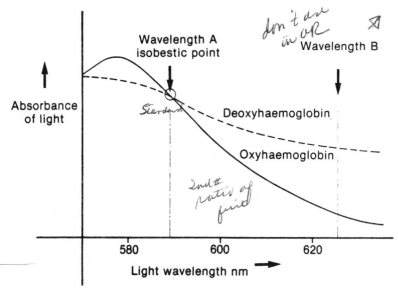

Figure 18.1 Variation of the absorption of light, according to its wavelength, by oxyhaemoglobin and deoxyhaemoglobin.

The graph shows the proportion of light absorbed at different wavelengths. The absorbances for oxyhaemoglobin and for deoxy-haemoglobin are generally different, but at point A the proportion of light absorbed is identical for the two compounds. This point is known as an isobestic point. In contrast, at the wavelength B on the right there is a wide difference between the light absorbed by the two forms of haemoglobin.

In a sample of blood, both forms of haemoglobin are present, so the absorbance at wavelength B lies somewhere between the two curves and this value can be compared with the value at wavelength A. At A the absorbance does not depend on the degree of oxygenation of the haemoglobin; thus A acts as a reference point to compensate for changes in absorbance due to changes in total haemoglobin concentration. The difference between the readings at A and B depends only on the ratio of oxyhaemoglobin to deoxyhaemoglobin and thus on the oxygen saturation. From readings of a sample at the two wavelengths A and B the oxygen saturation may be calculated. The actual calculations are quite complicated, but in modern oximeters the readings are usually computed electronically to give a direct reading of the saturation.

Figure 18.2 gives a simplified illustration of one type of oximeter. The blood sample to be analysed is drawn into the apparatus by a system of pumps to be diluted with a fluid and haemolysed before entering a cuvette. Light passes through filter A to provide a monochromatic beam of a single wavelength, representing for example that at the isobestic point A of Fig. 18.1. This light passes through the cuvette of haemolysed blood and is then detected by a photocell shown on the right. A second photocell acts as a reference detector picking up a fixed fraction of the light deflected by a beam splitter as shown, so that comparison of the output of the two detectors indicates the light absorbed. If a second filter B then replaces the first, the absorbance of light at another wavelength representing B on the graph (Fig. 18.1) can be measured and comparison of the results allows computation of the oxygen saturation of the blood. In addition, the total haemoglobin may be computed from the output of the photocells and the oxygen content estimated if required.

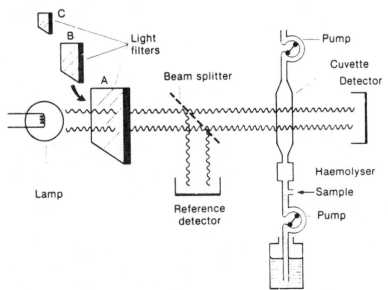

Figure 18.2   Principle of action of an oximeter.

CORRECTION FOR CARBOXYHAEMOGLOBIN

The effects of traces of carbon monoxide in the blood may be important, especially in cigarette smokers in whom up to 15% of haemoglobin can be in the form of carboxyhaemoglobin due to inhaled carbon monoxide. City dwellers may also have small amounts of carboxyhaemoglobin in their blood from carbon monoxide in the air.

The oximeter illustrated avoids errors from the presence of carboxyhaemoglobin by careful selection of the light wavelengths used in analysis and by electronic computation. In addition, carboxyhaemoglobin may be measured by using a series of light filters for differing frequencies of light; there are also oximeters using the same principle that provide a measurement of methaemoglobin in addition to carboxyhaemoglobin. In one model there is a series of filters mounted on a rotating disc, so that immediate measurement of all forms of haemoglobin can be obtained.

Techniques involving non-haemolysed whole blood are sometimes used and there are also oximeters that can be used on patients *in vivo*. In the latter instruments, the light source and detector are applied directly to a patient's forehead or earlobe and the reflected or transmitted light is again detected by means of photocells. These instruments are more difficult to calibrate accurately than the *in vitro* systems but do provide a system for continuous monitoring of oxygen saturation without the need for arterial puncture.

OXYGEN CONTENT

The measurement of oxyhaemoglobin gives an indication of the oxygen content of the blood, provided that the haemoglobin concentration and its oxygen combining power are known and that the oxygen in solution in the plasma is taken into account.

$$\text{Oxygen content} = \begin{array}{l} \text{(Blood oxygen saturation} \times \text{Haemoglobin} \\ \text{concentration} \times \text{Oxygen combining power} \\ \text{of haemoglobin)} + \text{oxygen in solution} \end{array}$$

It is more accurate, however, to measure the oxygen content of a blood sample directly, and an apparatus known as the 'Lex-O_2-Con' may be used.

The principle of the 'Lex-O_2-Con' is illustrated in Fig. 18.3. A carrier gas of 97% nitrogen, 2% hydrogen and 1% carbon monoxide passes first over a catalyst. This ensures that the hydrogen can combine with any traces of oxygen in the carrier stream to provide an oxygen-free carrier gas. The gas then bubbles through distilled water, contained in a circular glass tube known as the scrubber. The heparinised blood sample is

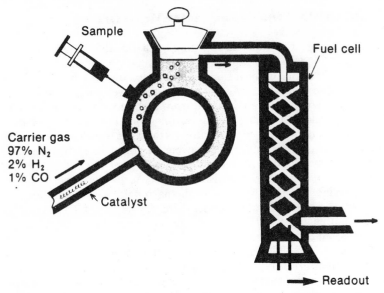

Figure 18.3 The 'Lex-O$_2$-Con' apparatus for the measurement of oxygen content.

injected from a syringe into this scrubber, where it is haemolysed by the distilled water and releases its oxygen into the water. The oxygen passes into the bubbles of carrier gas because the partial pressure of oxygen in these bubbles is zero. The carbon monoxide in the carrier gas combines with the haemoglobin released from the red cells and thus all the oxygen is displaced from the haemoglobin and enters the bubbles. The carrier gas with the oxygen in it then passes on to the cell shown on the right. This cell is very similar to the fuel cell to be considered later and in it oxygen molecules react at the cathode with the electrons (e) and water present to give hydroxyl ions.

$$O_2 + 4e + 2H_2O \rightarrow 4(OH)^-$$

In the cell, the total number of electrons used up is proportional to the total number of oxygen molecules reaching the cell from the sample.

These electrons may be measured by integrating electronically the current flowing through the cell as the oxygen is used up. Hence, a measure of the oxygen reaching the cell is obtained. Because the only source of oxygen is the sample, the readout gives the oxygen content of the sample of blood. The carrier gas, after passing through the fuel cell, bubbles out through a reservoir of distilled water to prevent any oxygen reaching the cell from the gas outlet.

Calibration is carried out with room air, and a chart is provided to correct the reading to a standard temperature and pressure.

The oxygen content of blood may also be measured traditionally by means of a Van Slyke apparatus. In this case, blood is haemolysed by an acid saponin mixture in a special apparatus and the contained gases extracted by a vacuum. Alkaline sodium dithyonite and alphahydroquinone is used to absorb the oxygen, and the oxygen content is calculated from the resultant fall in pressure in the gases extracted. The Van Slyke method requires a larger sample of blood and greater technical skill than with the 'Lex-O_2-Con' apparatus.

INTER-RELATION OF OXYGEN SATURATION AND OXYGEN TENSION

As an alternative to oxygen saturation measurement, the oxygen tension (Pao_2) of blood can be measured and there is a relatively consistent relationship between these measurements as shown by the oxygen dissociation curve (Fig. 18.4). Some instruments which measure Pao_2 also compute automatically the saturation after compensating for the effects of blood pH on the curve. This is not ideal because, in addition to pH, the curve is affected by other factors including the level of 2,3 diphosphoglycerate (DPG) in the red blood cells and direct measurement gives better accuracy than computation.

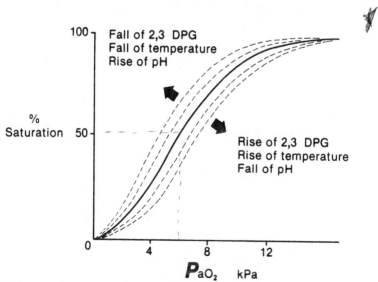

Figure 18.4   Oxygen dissociation curve.

At lower than normal levels of oxygen saturation, a small change of Pa_{O_2} is associated with a large change in saturation; hence oxygen saturation measurement may give a more accurate indication of a patient's oxygenation. In addition, of the two measurements, oxygen saturation is normally more relevant to patient care because it gives an index of the oxygen content of the blood and so of the oxygen which can be supplied to the tissues. Although Pa_{O_2} gives a measure of the partial pressure available for diffusion of the oxygen across membranes and capillaries, this is probably less important because oxygen diffuses readily across the pulmonary capillaries (Chapter 7), and because the mitochondria in the tissues can still use oxygen at tensions of a small fraction of a kilopascal (i.e. a few mmHg).

In respiratory study measurements, however, it is often necessary to measure high values of saturation or tension, and in this range a large change of Pa_{O_2} is associated with a range of 98–100% in saturation, so Pa_{O_2} measurement gives greater sensitivity.

MEASUREMENT OF OXYGEN TENSION

Blood gas analysers are available which allow measurement of the oxygen tension in a sample of blood in addition to measurement of the carbon dioxide tension and the pH of the sample. For oxygen tension measurements the apparatus incorporates an oxygen electrode, also known as a Clark electrode or a polarographic electrode. The principle of this electrode is illustrated in Fig. 18.5.

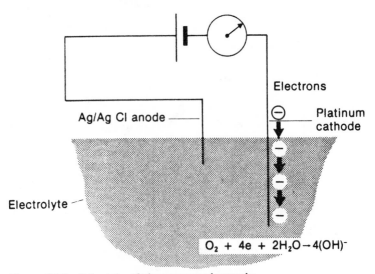

Electrons

Ag/Ag Cl anode

⊖ Platinum cathode

Electrolyte

$O_2 + 4e + 2H_2O \rightarrow 4(OH)^-$

Figure 18.5 Principle of the oxygen electrode.

A platinum cathode and a silver/silver chloride anode are shown in a solution of electrolyte such as potassium chloride. A voltage of 0·6 V is applied between the electrodes, and the current flow is measured. At the anode, electrons are provided by the reaction of silver with the chloride ions of the potassium chloride electrolyte to give silver chloride and electrons. At the cathode, oxygen combines with the electrons and water giving rise to hydroxyl ions.

$$O_2 + 4e + 2H_2O \rightarrow 4(OH)^-$$

The more oxygen available, the more electrons which can be taken up at the cathode and consequently the greater the current flow. So the current flow through the cell is dependent on the oxygen tension at the platinum cathode.

The reaction at the cathode is analogous to that occurring in cell mitochondria (Fig. 18.6). In this case the electrons produced in the Krebs cycle are transmitted down the chain of respiratory enzymes. In this respiratory chain, oxygen combines with the electrons to give rise to hydroxyl ions, while at the same time adenosine triphosphate (ATP) is formed from the diphosphate (ADP).

The platinum cathode of the oxygen electrode cannot be inserted direct into the blood because deposits of protein form and affect its

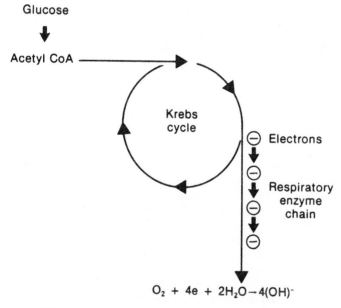

Figure 18.6 Krebs cycle and respiratory enzyme chain.

function and so, as shown in Fig. 18.7, there is a plastic membrane separating the blood from a small amount of electrolyte lying over the tip of the platinum cathode. The oxygen tension in the blood equilibrates with that in the electrolyte around the cathode and the reaction described above takes place, so that the current flow is dependent on the oxygen tension at the cathode.

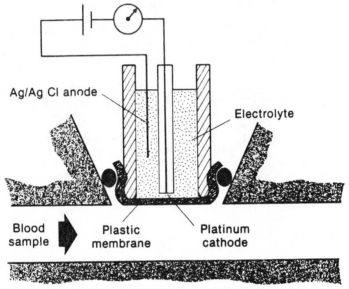

Figure 18.7   Oxygen electrode.

The oxygen electrode may be used with gases as well as with blood; two standard gas mixtures usually being provided for its calibration, one of which is oxygen-free and the other of known oxygen concentration. Temperature control is important, and the electrode must be carefully maintained at 37 °C.

The oxygen electrode must be kept scrupulously clean to prevent contamination, and its plastic membrane should be checked and replaced if punctured or if deposits of protein form on it. The arterial blood sample must be taken anaerobically and heparinised to prevent clotting, and analysis must be made promptly as the oxygen tension in an arterial blood sample falls steadily—particularly if it is maintained at room temperature. This fall in oxygen concentration is caused by the metabolism in the cells in the blood. If delay is unavoidable, then a correction factor is sometimes applied to allow for the time since the sample was taken. Alternatively, samples may be placed in a receptacle full of ice to reduce the rate of fall of the oxygen tension. These techniques are less accurate, however, than immediate analysis of the sample.

A miniature oxygen electrode mounted on the tip of a catheter has been produced to allow continuous measurement of the oxygen tension of blood inside the right heart. It readily shows changes in oxygen tension, but calibration is more difficult than with the conventional larger apparatus described.

False high readings may arise when some oxygen electrodes are used in the presence of halothane as the anaesthetic is reduced by the polarising voltage of 0·6 V. However, the use of an electrode membrane which is not readily permeable to halothane can minimise this problem.

THE FUEL CELL

Although the oxygen electrode may be used to measure the partial pressure of oxygen in gas mixtures, in practice several other types of apparatus are commonly used for this purpose. The fuel cell is considered first as it has some similarities to the oxygen electrode.

Figure 18.8 illustrates the similarity between the oxygen electrode and the fuel cell. In the oxygen electrode there is a platinum cathode, while in the fuel cell the cathode is in the form of a gold mesh. However, the reactions at the cathode are the same in both cases.

$$O_2 + 4e + 2H_2O \rightarrow 4(OH)^-$$

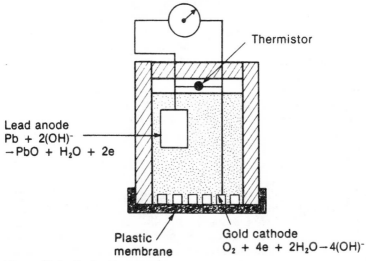

Thermistor

Lead anode
$Pb + 2(OH)^-$
$\rightarrow PbO + H_2O + 2e$

Plastic
membrane

Gold cathode
$O_2 + 4e + 2H_2O \rightarrow 4(OH)^-$

Figure 18.8   Fuel cell.

The current flow depends on the uptake of oxygen at the cathode and thus on the partial pressure of oxygen. The anodes, however, are different in the two cases. In the oxygen electrode there is a silver/silver chloride wire anode, but in the fuel cell the anode is made of lead and the electrolyte is potassium hydroxide. At the anode, electrons are produced by the combination of hydroxyl ions from the potassium hydroxide electrolyte with the lead as follows:

$$Pb + 2(OH)^- \rightarrow PbO + H_2O + 2e$$

Unlike the oxygen electrode, no battery is needed in the fuel cell because the cell produces a voltage, thus acting as a battery. Like other batteries it eventually expires, and the actual life of the cell depends on the period during which it is exposed to oxygen, which may be several months. As with the polarographic electrode, temperature affects the output of a fuel cell but temperature compensation may be achieved by means of a thermistor placed within the fuel cell as shown in Fig. 18.8.

The cell is connected to a small indicator unit to provide a portable and reliable oxygen analyser for clinical and bedside use. A typical cell has a response time of about 30 s. Although this type of analyser is not expensive, the detector cells have a limited life span and do need replacing. Additionally, if a standard cell is used in the presence of nitrous oxide, nitrous oxide diffuses into the cell. At the anode the nitrous oxide reacts with the lead in the presence of the electrolyte to produce nitrogen, which alters the pressure inside the cell and may damage it. Nevertheless, there are special cells designed for use in anaesthetic practice when nitrous oxide might be present, but they have a shorter life span than the standard ones.

Portable oxygen sensors based on the oxygen electrode are also available. In this type, the electrolyte is in the form of a gel but this and the membrane deteriorate fairly rapidly with use and must be replaced. This type has the advantage of a faster response but, on the other hand, the fuel cell oxygen sensor requires less maintenance and has a guaranteed life span.

Many of these oxygen sensors incorporate a visual or audible alarm to alert staff should monitored oxygen levels deviate from preset limits.

PARAMAGNETIC ANALYSER

Another technique for measuring the concentration of oxygen in gases is based on the fact that O_2 is paramagnetic and is, therefore, attracted into a magnetic field (Fig. 18.9). On the other hand, most other gases, e.g. nitrogen, are weakly diamagnetic and are repelled from a magnetic field. The paramagnetic property of oxygen arises from its molecular structure, being due to the fact that the electrons in the outer shell of an

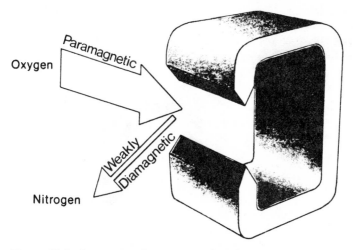

Figure 18.9   Paramagnetic property of oxygen.

oxygen molecule are unpaired. In the paramagnetic oxygen analyser a gas-tight chamber or cell containing two glass spheres connected in a dumb-bell arrangement is placed in a non-uniform magnetic field (Fig. 18.10).

The dumb-bell is suspended on a filament so that it can rotate, and the glass spheres are filled with nitrogen. The zero reading of the instrument may be adjusted by filling the chamber around the spheres with nitrogen. The orientation of the dumb-bell in the magnetic field is determined by the tension of the suspending filament. Suppose oxygen is added to the cell; the oxygen is attracted into the magnetic field and tends to displace the glass spheres containing nitrogen. The dumb-bell therefore rotates until the force of this displacement is balanced by the tension of the filament. This deflection of the dumb-bell may be measured by attaching a mirror to the dumb-bell suspension as illustrated, so that the position of a light beam reflected from this mirror provides an indication on a scale. The scale can be calibrated to give the percentage of oxygen present in the measuring cell.

Figure 18.11 illustrates an alternative paramagnetic technique. The reflected light beam falls on two photocells, and the output from these is compared and used to generate a current that flows through a coil around the dumb-bell. This current produces an opposing magnetic field which keeps the dumb-bell in its resting position. The amount of current required to keep the dumb-bell steady gives a measure of the concentration of oxygen in the cell. Such null deflection analysers give very accurate results, typically 0.1%.

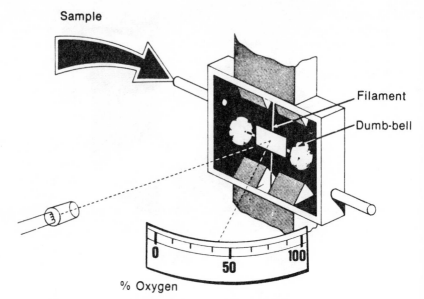

Figure 18.10   Simple paramagnetic oxygen analyser.

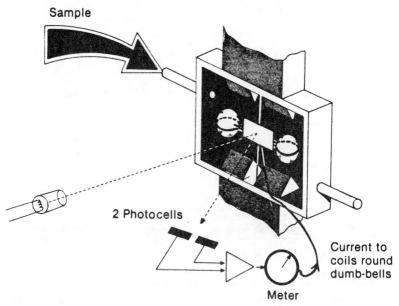

Figure 18.11   Null deflection type of paramagnetic oxygen analyser.

As with all instruments, paramagnetic analysers must be calibrated before use, e.g. with 100% nitrogen and 100% oxygen in the cell. A practical point to note is that, as with many other gas analysers, water vapour biases the reading. For the most accurate analysis, therefore, sample gases should be dried by passage through silica gel before they enter the analysis cell.

ANALYSIS OF GASES CONTAINING WATER VAPOUR

If water vapour is added to a previously dry mixture of gases, the concentration of the other components will be reduced because of the partial pressure of the water vapour. This may sometimes be important. Consider, for example, the measurement of the oxygen consumption of a patient who inspires dry gas but who expires humidified gas. Unless a correction is made for the addition of water vapour to the gas mixture, the oxygen consumption result will be inaccurate.

If humidified gases are analysed at 37 °C, the oxygen concentration in the dry mixture can be calculated from the partial pressure of oxygen, the barometric pressure and the saturated vapour pressure (s.v.p.) of water at 37 °C.

$$\text{Concentration} = \frac{\text{Partial pressure (humidified at 37 °C)}}{\text{Barometric pressure} - \text{s.v.p. of water at 37 °C}}$$

As an alternative to calculation the water vapour may be removed by passing the gases through a drying agent such as silica gel before analysis.

CONTINUOUS *IN VIVO* OXYGEN MEASUREMENT

Most of the instruments described are used for intermittent sampling of arterial blood, mixed venous blood or of gas samples, but in the seriously ill patient a single reading may not be representative and continuous recording of the oxygen tension may be desirable. Several techniques are available to meet this need. Thus, a miniaturised polarographic electrode may be used in a catheter in an artery, or in a central venous line. The mass spectrometer (Chapter 20) has been used to give a continuous measurement of oxygen tension using an intravascular catheter.

Unfortunately, calibration remains a problem with most of the continuous recording systems, Consequently, serial sampling from an indwelling arterial catheter or from a central venous line or from end-tidal air samples is still used in the management of the seriously ill patient. However, an alternative technique, that of transcutaneous oxygen measurement, is useful as a complement to blood gas analysis, particularly in neonates.

TRANSCUTANEOUS OXYGEN MEASUREMENT

At room temperature, some of the oxygen supplied by the peripheral circulation is used in metabolism by the tissues underlying the skin surface. However, if the skin is heated locally to 43–44 °C, the peripheral blood vessels dilate and the oxygen used in the skin is negligible compared to the total oxygen present in the capillaries. The skin and capillary tensions then approximate to the arterial tension. Oxygen diffuses through the skin. Therefore, an oxygen electrode placed on the skin surface can be used to measure the arterial oxygen tension.

A cross-section through a typical transcutaneous oxygen electrode is shown in Fig. 18.12. A platinum or gold cathode is surrounded by a silver anode, both electrodes being in contact with a thin layer of electrolyte retained by a membrane permeable to oxygen. The complete system forms a polarographic electrode, the operation of which has been described above. The skin temperature is measured by a thermistor, and this is used to control the heater power. Failure of the

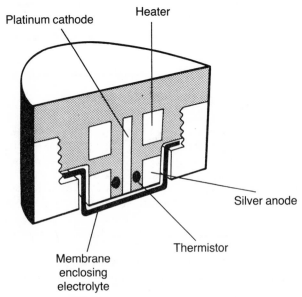

Figure 18.12 Transcutaneous oxygen electrode.

temperature control system is dangerous as burns can occur if the skin temperature rises above 45 °C (Chapter 9). Consequently, a second thermistor is used to switch off the heater if the primary heating circuit fails. Nevertheless, to avoid damage to the skin and to maintain accurate operation, the electrode position should be changed every 3–6 hours.

Transcutaneous oxygen measurement is less accurate than analysis of blood samples, but as it is continuous it can be used to complement the latter. Inaccuracies arise from the metabolism of oxygen as it diffuses through the skin, the shift in the oxygen dissociation curve with temperature, and the reduction of skin perfusion due to oedema resulting from the high operating temperature of the electrode. Inaccurate low readings also arise if cardiac output falls, as skin perfusion is then reduced. The response time of the transcutaneous electrode is rather slow. The electrode responds in about 1 min to changes of oxygen tension at its surface. Furthermore, the changes in cutaneous tension in turn lag behind changes in arterial blood tension.

CHAPTER 19

Measurement of pH and CO_2

THE CONCEPT OF pH

pH is a measure of the hydrogen ion activity in a liquid. Hydrogen ion activity is not exactly the same as hydrogen ion concentration but, for practical purposes in the clinical situation, these may be regarded as equivalent. pH is defined as the negative logarithm to the base 10 of the hydrogen ion concentration. Hence a hydrogen ion concentration of 10^{-9} mol litre^{-1} would be a pH of 9. Each decrease of one pH unit is equivalent to a ten-fold increase in hydrogen ion concentration.

pH of 9 = 1 nmol litre^{-1} (H$^+$) or 10^{-9} mol litre^{-1}

pH of 8 = 10 nmol litre^{-1} (H$^+$) or 10^{-8} mol litre^{-1}

pH of 7 = 100 nmol litre^{-1} (H$^+$) or 10^{-7} mol litre^{-1}

pH of 7·4 = 40 nmol litre^{-1} (H$^+$) or $10^{-7·4}$ mol litre^{-1}

The normal blood pH is 7·4 at 37 °C, and this is equal to 40 nmol litre^{-1} hydrogen ion concentration.

In some laboratories nanomoles per litre is replacing pH as a measure of the hydrogen ion concentration, but pH remains the common unit and is used in this chapter. Full details of the clinical problems surrounding the maintenance of the acid-base state of the blood are beyond the scope of this book.

MEASUREMENT OF pH

The pH electrode assembly (Fig. 19.1), contains a capillary tube of pH-sensitive glass across which a potential develops which depends on the difference in pH across the glass. The pH outside the capillary is maintained at a constant value by a buffer solution, so that the potential across the glass is proportional to the pH of the blood inside the capillary. In order to measure this potential it is necessary to make electrical contact with the blood and with the buffer solution by means of special stable electrodes. These are unaffected by changes in the solutions with which they are in contact. Such electrodes consist of a metal in contact with the metal chloride which, in turn, is in contact with a saturated solution of chloride ions. Mercury or silver are suitable metals and they are incorporated into electrodes as shown in Fig. 19.1. On the left of the diagram is the pH-sensitive glass capillary surrounded by a jacket of buffer solution containing a silver/silver chloride electrode, which provides an electrical connection to the buffer solution. On the right of the diagram is a mercury reference electrode, linked by a porous plug to a potassium chloride solution. The blood is in contact with this potassium chloride bridge via a membrane; thus there is a complete electrical circuit. The potential difference between the electrodes is displayed on the meter and, as the potential varies with the pH of the blood, the meter can be calibrated to give a direct reading of pH.

There is an automatic suction control to draw the blood through the capillary. Automatic systems of this type are often arranged as part of a comprehensive electrode system which measures oxygen and carbon dioxide tensions in addition to pH. Normally, the blood sample is an arterial one taken anaerobically, heparinised and analysed promptly.

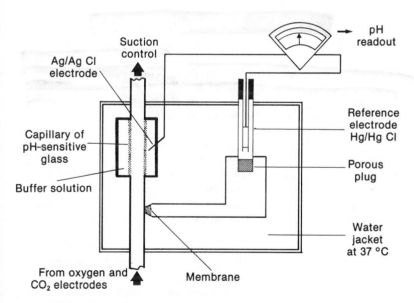

Figure 19.1   pH electrode system.

Temperature control is very important when measuring blood pH and, as seen in Fig. 19.1, a water jacket at 37 °C surrounds both the pH-sensitive capillary with its buffer jacket and the reference electrode. Dissociation of acids or bases increases with temperature and thus alterations in temperature change the pH. If the blood sample is not at 37 °C, e.g. if it is from a patient under hypothermic conditions, then a correction factor may be used to indicate the true pH in the patient.

Before use, pH meters must be calibrated with two buffer solutions, each containing a fixed composition of two phosphate buffers whose pH values have been decided by international agreement. In addition to their use for the measurement of pH in arterial blood, these systems may also be used to indicate the pH of urine or of cerebrospinal fluid.

MEASUREMENT OF P_{CO_2}

The carbon dioxide tension in arterial blood (Pa_{CO_2}) is a valuable measurement of the adequacy of a patient's ventilation, the tension of carbon dioxide in blood being the partial pressure of carbon dioxide which would be in equilibrium with the blood. The normal Pa_{CO_2} is 5·3 kPa or 40 mmHg.

The main methods of measuring the P_{CO_2} of liquids are based on pH measurement because carbon dioxide reacts with water to produce hydrogen ions by the reversible reactions:

$$CO_2 + H_2O \rightleftharpoons H_2CO_3 \rightleftharpoons H^+ + (HCO_3)^-$$

The first reaction gives carbonic acid which, in turn, dissociates to produce hydrogen ions and bicarbonate ions. The carbon dioxide tension is therefore related to the hydrogen ion concentration; hence the logarithm of the carbon dioxide tension is related to the logarithm of the hydrogen ion concentration and thus to the pH. This interrelationship of P_{CO_2} to pH is used in the Severinghaus CO_2 electrode.

Figure 19.2 illustrates the Severinghaus CO_2 electrode which provides a direct method of P_{CO_2} measurement from the pH change. The apparatus incorporates pH-sensitive glass, shown in the centre, with electrodes on each side of it. The pH-sensitive glass is in contact with a thin film of sodium bicarbonate solution in a nylon mesh which is fixed over the glass tip by means of an O-ring. The liquid to be tested, usually arterial blood, is separated from the nylon mesh and bicarbonate by a plastic membrane that is permeable to carbon dioxide and which is also attached by means of an O-ring. At the tip of the electrode, carbon dioxide diffuses through the plastic membrane into the mesh impregnated with the bicarbonate solution and combines with the water present as described above. The presence of the resulting hydrogen ions lowers the pH of the sodium bicarbonate solution. This pH change is

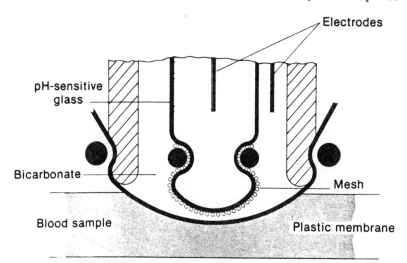

Figure 19.2   Severinghaus CO_2 electrode.

measured by the glass electrode. As pH changes linearly with the logarithm of the carbon dioxide tension, the machine can be calibrated in terms of P_{CO_2}.

If a small hole occurs in the plastic membrane, it then ceases to be a semipermeable membrane for carbon dioxide and the apparatus becomes inaccurate, but with care an accuracy of about 130 Pa (1 mmHg) may be obtained. The response time of the electrode is about two to three minutes, being slower than the pH electrode because the carbon dioxide requires time to diffuse through the plastic membrane. As with the pH electrode, the carbon dioxide electrode must be maintained at 37 °C, and be calibrated before use with standard gas mixtures containing known concentrations of carbon dioxide. As the meter gives readings as tensions or partial pressures in kPa or mmHg, it is necessary to convert the concentrations of the calibration mixtures to partial pressures. It must be remembered that, although the partial pressure of a dry gas can be obtained by multiplying the concentration by the barometric pressure, for humidified gas the saturated vapour pressure of water must be subtracted from the barometric pressure.

In the blood gas apparatus, the calibration mixtures are fully humidified on their way to the electrode. An example of the calibration for carbon dioxide is as follows. If the barometric pressure is 100 kPa and the saturated vapour pressure of water at 37 °C is 6·3 kPa, a 5% carbon dioxide concentration calibrating gas gives a partial pressure of 4·7 kPa.

$$\frac{5}{100} \times (100 - 6 \cdot 3)\,\text{kPa} = 4 \cdot 7\,\text{kPa}$$

The calculation in terms of millimetres of mercury is

$$\frac{5}{100} \times (750 - 47) \text{ mmHg} = 35 \text{ mmHg}$$

For a 5% carbon dioxide calibrating gas in this example, therefore, the meter calibration must be set to read 4·7 kPa or 35 mmHg, depending on the machine.

The Pa_{CO_2} provides a measure of the carbon dioxide in solution in the blood. However, this is only a small part of the total carbon dioxide, which also includes carbon dioxide combined as bicarbonate and carbonate and in association with haemoglobin. However, in anaesthesia the Pa_{CO_2} is the most important measurement.

The carbon dioxide electrode suffices for intermittent measurement of blood and gas samples but in very ill patients it could be advantageous to have a continuous *in vivo* recording. New techniques with intravascular catheters are being developed to facilitate this based on the use of mass spectrometry (Chapter 20).

In the same way that transcutaneous oxygen electrodes can be used to measure oxygen tensions (Chapter 18), transcutaneous carbon dioxide electrodes are available which use a similar temperature control system but which use an electrode of the Severinghaus type.

INFRA-RED ANALYSER

The infra-red analyser is another system which allows the continuous analysis of gas samples.

Gases that have two or more different atoms in the molecule absorb infra-red radiation. The proportion of radiation absorbed depends on its wavelength as shown in Fig. 19.3. Each gas absorbs radiation at characteristic frequencies. By careful choice of an infra-red wavelength and by taking into account the potential components of the gas mixtures it is possible to avoid interference from other gases. In the example, analysis of a carbon dioxide mixture from absorption of infra-red radiation with a wavelength of 4·28 μm should avoid interference from the presence of nitrous oxide.

The principle of an infra-red analyser is shown in Fig. 19.4. The infra-red radiation is emitted by a hot wire and the particular frequency required is obtained by passing the radiation through a filter. An instrument designed to analyse carbon dioxide concentrations only could have a filter to give a fixed wavelength, whereas more versatile instruments incorporate variable filters so radiation wavelengths appropriate to a variety of different gases may be selected.

From the filter the radiation passes through a chamber containing the sample to be analysed. The chamber has windows made from a crystal of sodium chloride or sodium bromide, because glass absorbs

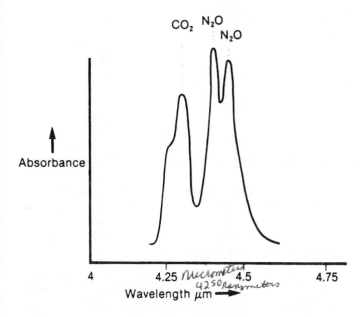

Figure 19.3   Absorption of infra-red radiation by carbon dioxide and nitrous oxide depends on the wavelength of the radiation.

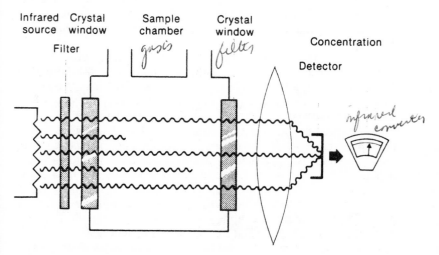

Figure 19.4   Infra-red analyser.

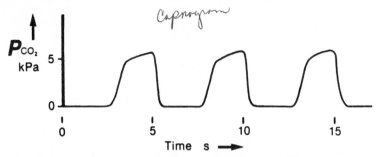

Figure 19.5   Recording of expired carbon dioxide tension.

infra-red radiation. From the chamber the radiation is focused on a semiconductor detector. The greater the absorption of infra-red radiation by the carbon dioxide in the sample the less the radiation monitored by the detector. Consequently, it is possible to process the detector output electronically to indicate the concentration of the gas present.

In practice, it is customary to fill the sample chamber initially with a carbon dioxide-free mixture to check that the meter reads zero correctly; some instruments use a second beam, which passes through a reference chamber, to compensate for variation in the output of the infra-red source and variation in transmission of the optical components in the system.

The sample chamber is small, so that continuous analysis is possible if a continuous flow of gas is passed through the apparatus. In this way a continuous recording of carbon dioxide concentrations in a respiratory circuit may be obtained (Fig. 19.5). The response time is about 100 ms and the tracing enables estimation of the end-tidal carbon dioxide concentration.

Because of the short response time the technique gives prompt warning of mishaps during anaesthesia. Thus, the end-tidal carbon dioxide drops rapidly towards zero in a patient with an air embolus. Alternatively, a rise of end-tidal carbon dioxide may indicate inadequate ventilation or the onset of malignant hyperpyrexia.

The carbon dioxide electrode and the infra-red analyser are the principal techniques for analysing carbon dioxide, but gas chromatography and mass spectrometry (Chapter 20), and the interferometer (Chapter 13) may also be used. In addition to physical methods to measure carbon dioxide in a gas mixture, a chemical method may be used. One such technique is based on the reduction in the volume or pressure of a gas sample after the carbon dioxide in it has been absorbed by an alkaline solution such as potassium hydroxide. These techniques were commonly used in the past and may still be used for reference purposes in the laboratory.

Gas Chromatography and Mass Spectrometry

Gas chromatography and mass spectrometry are two versatile techniques which may be used to identify or measure the concentrations of anaesthetic agents or other gases present in a mixture. Although both procedures can be used in the operating theatre, they are more commonly employed as research techniques in the laboratory.

GAS CHROMATOGRAPHY

Chromatography literally means to write in colour. As shown on the left of Fig. 20.1, if a mixture of dyes is poured into a column packed with calcium carbonate the pigments separate out as coloured bands at different levels, and the components may eventually be collected from the outlet. Chromatography is now a general term for analytical procedures that separate a mixture into its components as the mixture passes through a column. The calcium carbonate in the example forms a stationary phase which retains the individual dyes for different periods, hence separating the mixture. The dye mixture passing through forms a mobile phase.

In a gas chromatograph there is also a stationary phase and a mobile phase as shown on the right of Fig. 20.1. The stationary phase, with which the chromatographic column is packed, usually consists of a support material such as silica-alumina in the form of very small particles which is then coated with polyethylene glycol or silicone oil. Through the column containing this packing passes a flow of carrier gas, such as nitrogen, argon or helium, which forms the mobile phase. The gas mixture to be analysed is injected into the inlet of the column into the mobile phase and the components pass through the column at a speed which depends on their differential solubility between the two

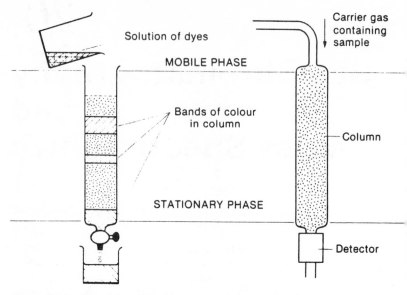

Figure 20.1   Principle of liquid and gas chromatography.

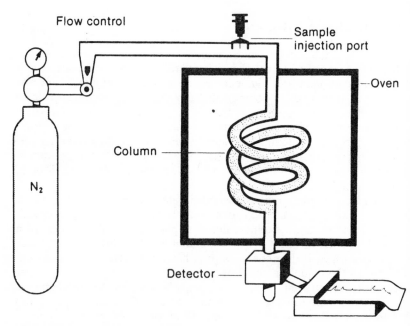

Figure 20.2   Gas chromatograph.

phases. The term gas liquid chromatography (g.l.c.) is sometimes used for this system. As solubility is temperature dependent, the column is maintained at a preset temperature in a thermostatically controlled oven (Fig. 20.2).

As shown in Fig. 20.2, the column is usually a long coiled glass tube containing the packing material. The added length of the column gives better separation of the components. A flow control provides a steady stream of carrier gas, in this case nitrogen, through the column and an injection port for the sample is provided at the column inlet.

DETECTOR

At the outlet of the column a detector is connected to a recorder to monitor the appearance of the sample components. A typical example is the flame ionisation detector (Fig. 20.3).

In the flame ionisation detector, hydrogen and air are added to the gas to produce a flame at the end of the column. The constituents of a flame are ionised particles. If a polarising voltage is applied across the flame by two electrodes, a current is produced. The magnitude of this current depends on the particular charged particles in the flame. If a sample of organic vapour is present in the carrier gas, it is ionised in the

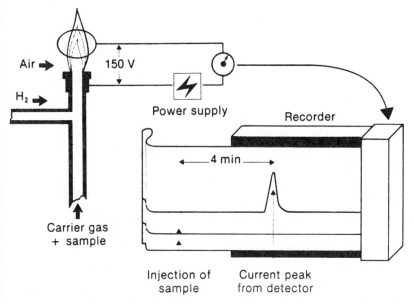

Figure 20.3   Flame ionisation detector.

flame and increases the current flow. The size of the current increase gives a measure of the amount of organic vapour present.

As shown in Fig. 20.3 the gas chromatograph recording appears in the form of a peak as each component of the sample passes out of the column.

There are two main alternatives to the flame ionisation detector: the thermal conductivity detector (also called the katharometer), and the electron capture detector. The katharometer (Fig. 20.4A) measures changes in thermal conductivity of the gas by the change in the

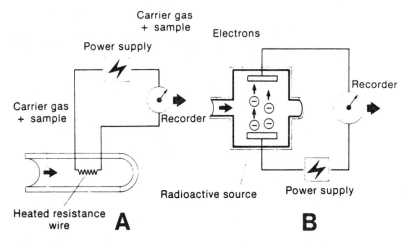

Figure 20.4 (A) Thermal conductivity detector. (B) Electron capture detector.

electrical resistance of a heated wire placed in the gas flow. It is more suitable than the flame ionisation detector for the analysis of inorganic gases such as nitrous oxide or oxygen.

Halogenated compounds can be detected with greater sensitivity by an electron capture detector (Fig. 20.4B). In this detector a polarising voltage is applied across an ionisation chamber in which electrons, released from a radioactive cathode, constitute a flow of current to the anode. Halogenated compounds from a sample capture these electrons and reduce the current flow.

None of these detectors allows the absolute identification of the unknown components of a sample, and it is necessary to have preliminary knowledge of the types of substances present and to apply this knowledge to the analysis of the recorder tracing.

ANALYSIS OF THE RECORDER TRACING

As different pure substances progress at characteristic rates through the separation column, the time between the injection of a sample and its appearance at the detector may be used to identify the component. This period is known as the retention time, and is four minutes in the example given in Fig. 20.3. In practice, a mixture of substances may be present in the sample and several peaks, each with its characteristic retention time, appear on the tracing (Fig. 20.5).

Provided that the nature of the sample components is known, calibration samples of known quantities of these can be injected as shown in Fig. 20.5. These allow confirmation of the presence of the substances and the comparison of the peak heights (or, more accurately, peak areas) permits calculation of the quantity of each substance.

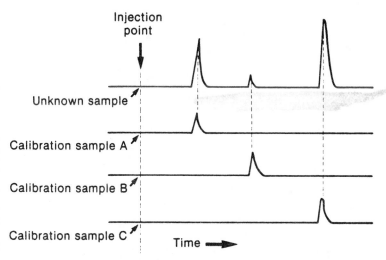

Figure 20.5 Comparison of the chromatograph peaks from a sample with those of calibration samples.

INTRODUCTION OF A SAMPLE INTO THE GAS CHROMATOGRAPH

The sample is usually injected as rapidly as possible from an appropriate gas tight syringe through a rubber septum. If the port at which the injection is made is heated, then injected liquids are vaporised and the volatile components may be analysed. In this way, even blood samples can be analysed direct in a gas chromatograph.

As an alternative to the use of a gas tight syringe, gas samples may be introduced by a special gas sampling valve.

CLINICAL USE

Gas chromatography is used for measuring volatile anaesthetic agents, both for anaesthetic concentrations and for trace levels in the theatre atmosphere.

The gas chromatograph is also of use in the measurement of barbiturates, phenothiazines, benzodiazepines, steroids and catecholamines, special techniques being used to convert these into volatile compounds.

Gas chromatography has several advantages in that it allows the identification and measurement of very low concentrations of drugs, even in a mixture of compounds; and also the technique is versatile in that both liquids and gas samples can be analysed. However, its disadvantages are that continuous analysis is impossible and details of the sample constituents must be available.

THE MASS SPECTROMETER

References are made in other chapters to the use of the mass spectrometer to measure water vapour, oxygen and carbon dioxide, but this instrument can also be used to separate and measure other gases. The principle of action is illustrated in Fig. 20.6. On the left, a few molecules of the sample pass through a molecular leak into an ionisation chamber where they are bombarded by a beam of electrons passing from the hot cathode to the anode. This beam of electrons is produced in the same way as in the oscilloscope. When the molecules of the gas are hit by the electrons, some of them become charged ions which are then accelerated out of the chamber in a narrow beam by means of the acceleration and focusing plates. The stream of ions then passes through a strong magnetic field. The charged ions are deflected in an arc by the magnetic field, the amount of deflection depending on their mass—the lighter ions being deflected most, the heaviest least. In Fig. 20.6 four different ion streams are shown, only one of which passes through a small slit to be picked up by a detector. The signal from the detector depends on the number of ions reaching it, and this signal can be amplified and displayed. By varying the voltage on the acceleration and focusing plates, the position and speed of the beam may be altered and streams of ions of different masses detected.

Another method of picking out specific streams of ions is shown in

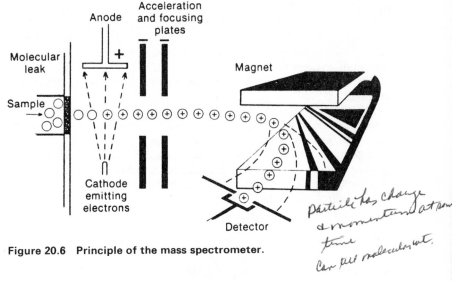

Figure 20.6 Principle of the mass spectrometer.

[handwritten notes:] particle has charge & momentum at same time. Can tell molecular wt.

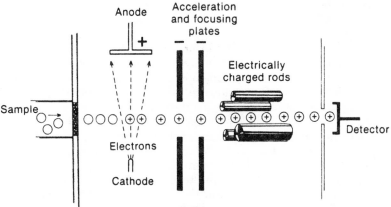

Figure 20.7 Quadrupole mass spectrometer.

Fig. 20.7, and uses four electrically charged rods in place of the magnetic field. The potentials on the rods are varied so that the ions oscillate between them as they travel. It can be arranged so that only ions of a specific mass are able to travel the length of the rods without being removed from the stream. This system is known as a quadrupole mass spectrometer.

Although the mass spectrometer identifies compounds in terms of their mass numbers, some compounds fragment during the ionisation process. Thus, it is sometimes possible, by choosing an appropriate breakdown compound for detection and measurement, to distinguish between compounds of the same mass number. An example of this in anaesthesia arises in the analysis of nitrous oxide in the presence of carbon dioxide, as both have the same mass number of 44. The nitrous oxide fragments on ionisation, and the smaller nitric oxide fragment can be identified.

The mass spectrometer is a versatile instrument which may be used to detect and measure many different gases and compounds. Because of its ability to identify compounds through their mass numbers and to analyse minute samples it is sometimes used in isotope studies and is also used in conjunction with gas chromatography for further analysis and measurement of the separated components issuing from the chromatograph column.

OTHER TECHNIQUES OF VAPOUR ANALYSIS

Although the gas chromatograph and mass spectrometer are highly versatile instruments they are bulky and less convenient for use in areas outside the laboratory. Anaesthetists often use alternative techniques in such areas.

The solubility of anaesthetic agents in oil is the principle used by one form of anaesthetic vapour analyser known commercially as the 'Emma'. When an electric potential is applied across a crystal of quartz it contracts slightly, this is known as the piezoelectric effect. The crystal can be made to oscillate at its resonant frequency by a suitable alternating potential, and the vapour analyser uses such a crystal to which a thin oily coating has been applied. Anaesthetic agents dissolve in the oily coating and alter the resonant frequency of the crystal. From Henry's law, the quantity of vapour which dissolves is proportional to the partial pressure of the vapour, so the change in resonant frequency of the crystal can be measured electronically and displayed as anaesthetic concentration. The analyser is not specific to individual anaesthetic agents, and responds to a limited extent to water vapour. Nevertheless, the analyser forms a useful instrument for measuring anaesthetic vapour concentration; for example, in the closed circuit.

Anaesthetic vapour concentration can also be measured by the infra-red analyser described in Chapter 19 (Fig. 19.4), and halothane concentration by an ultra-violet analyser which works on a similar principle. The ultra-violet light is produced by a lamp containing mercury vapour. The light beam is split into two and passes through a sample cell and a reference cell via their quartz windows. The ultra-violet light from each cell falls on a photodetector tube. Halothane absorbs ultra-violet light. Hence, the light reaching the photodetector tube beyond the sample cell varies in intensity with the concentration of halothane in the cell.

Another technique for the measurement of anaesthetic vapours is the use of the interferometer. This is appropriate for analysis of individual samples and for checking the performance of vaporisers; its principle is described in Chapter 13.

21

Gas Supplies

The maintenance of medical gas supplies and the care of medical gas cylinders are usually under the supervision of a hospital's engineering and technical staff but anaesthetists using gas supplies and cylinders should have some knowledge of the topic.

GAS CYLINDERS

Most large gas cylinders are stored in the upright position, but some smaller ones and Entonox cylinders are stored horizontally. Cylinders should be stored indoors, protected from the weather and should not be subjected to extremes of heat or cold; they should also be used in rotation with separate areas for full and empty cylinders. Cylinders containing medical gases should be stored separately from other types and separate areas should be allocated for flammable gases such as cyclopropane. To withstand the high pressure the cylinders are made of high carbon steel, manganese steel or aluminium alloy, although even these steels will not prevent the risk of an explosion if the cylinder is dropped on a hard concrete surface.

Cylinders are identified not only by clear written labels but also by colour coding. In the international standard of colours for medical gases (ISO/R32), oxygen cylinders are black with a white top, nitrous oxide blue, cyclopropane orange, and carbon dioxide grey. The international system is adopted in the UK, but in the USA and some other countries different colour codes may still be seen.

Around the neck of a cylinder in the UK is a plastic disc, the colour and shape of which denotes the year when the cylinder was last examined (Fig. 21.1). The manufacturers regularly examine and test cylinders, including an internal examination with an endoscope, so that any faulty ones may be withdrawn from use. The interval between testing varies but is usually from five to ten years, depending on the gas.

On the top of the cylinder is an identification label summarising safety precautions. There are risks of explosion and fire if oil or other flammable liquids are allowed to come into contact with high pressure

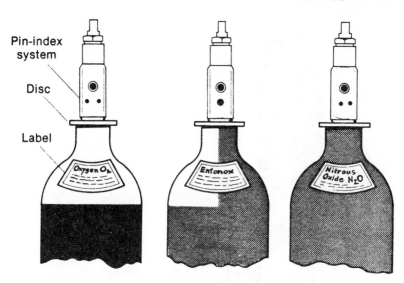

Pin-index
system

Disc

Label

Oxygen O₂

Entonox

Nitrous
Oxide N₂O

Figure 21.1   Gas cylinders—pin-index system and test discs.

oxygen, nitrous oxide or Entonox, since all these gases support combustion.

Heating cylinders is dangerous because it increases the pressure inside them. A moderate rise in temperature as occurs in summer, however, is not important because the cylinders are made to withstand pressures well above their normal working range. The actual value of the pressure that the cylinder can withstand depends on its type and on the gas concerned, but it is usually 65 to 70% above the working pressure.

The cylinder valve should be opened momentarily before attaching the cylinder to the anaesthetic machine. This blows out any dust or other material which might be lodged in the outlet of the cylinder and which might otherwise enter the anaesthetic apparatus.

After attaching the cylinder, it is recommended that the valve is opened slowly to avoid adiabatic heating (Chapter 4). It should be opened two full turns because valves that are only partially open may restrict the flow of gas when the cylinder pressure falls. Finally, only moderate force should be used when closing the valve as otherwise the valve seating can be damaged.

Figure 21.2 shows a detailed section of the valve seating. A spindle with a plastic facing is screwed up and down against the valve seating. This plastic facing may be damaged if the valve is closed too tightly. Around the spindle, below the nut at the top, is a compressible plastic

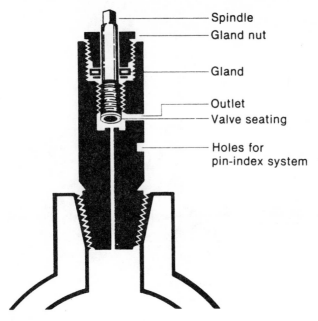

Figure 21.2   Section through a cylinder valve.

washer known as a gland which prevents any leak while still allowing the spindle to turn.

On the face of the valve block are holes which correspond with pins on the yoke of the anaesthetic machine. This arrangement comprises the pin-index system, a device to prevent connection of the wrong cylinder to the yoke of the anaesthetic machine. There is a special pin-index configuration for each medical gas; Fig. 21.1 illustrates the configuration for oxygen, Entonox and nitrous oxide.

Instead of the pin-index system there are two other systems used on the larger cylinders. A large bull-nose cylinder typically used for oxygen, is shown on the left of Fig. 21.3. It has a right-hand internal thread on the cylinder outlet identical to that used for the large air, helium and nitrogen cylinders. For hydrogen and other flammable gases, however, the cylinder outlet has a left-hand thread giving some protection against wrong connection.

On the right of the diagram, the valve of a large nitrous oxide cylinder is shown. Notice that it has an external thread connection and may also have a handwheel. Similar handwheel cylinders may also be seen outside the UK, often containing other gases but with different patterns of thread. These systems, however, give only partial protection against incorrect connection.

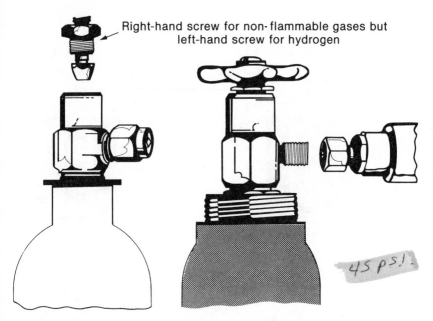

Right-hand screw for non-flammable gases but left-hand screw for hydrogen

45 p.s.i.

Figure 21.3   Thread patterns for larger cylinders.

PIPED OXYGEN   SUPPLIES

To avoid frequent changing of cylinders of gases in busy theatres and for reasons of economy, gases may be piped from a remote storage area. The commonest piped supply is that of oxygen and the central source may be a cylinder manifold or a liquid oxygen tank. However, in either case the oxygen storage area should be outside or not in direct communication with the main hospital buildings, because there is an increased fire risk in all areas where it is stored.

In many hospitals the oxygen supply is in the form of a cylinder manifold; Fig. 21.4 shows such a manifold with its cylinders attached. The six cylinders illustrated are divided into two banks of three and are so arranged that each bank contains the correct number of cylinders to last for a minimum of two days' normal use; hence the size of the manifold varies according to the size of the hospital.

In Figure 21.4 the three cylinders on the left are in use, the three on the right in reserve. Each cylinder is connected by a coiled tube to a pipeline which passes to a central control box. The high pressure gauges on the left and right sides of the box indicate the contents of the cylinder banks, and below are high pressure reducing valves which lower the cylinder pressure of 137 bar to about 10 bar. The changeover valve,

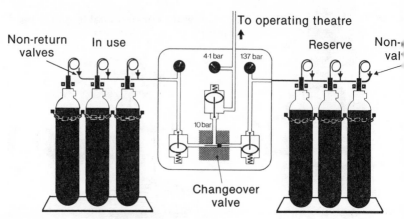

Figure 21.4 Oxygen manifold.

shown in the centre, has a pneumatic shuttle mechanism which is held at its extreme positions magnetically, but which switches automatically from the in-use bank to the reserve bank of cylinders when the pressure falls to a certain value. Electric warning devices are provided, and are activated by the changeover shuttle valve. These may switch on warning lights or audible alarms, both at the control box and at remote points to indicate the need to change the empty bank of cylinders. A secondary fail-safe bypass system with additional cylinders may also be present to ensure a continued supply even in the event of a changeover failure.

At the outlet of the shuttle valve is a second-stage reducing valve to reduce the pressure from 10 bar to 4·1 bar, the normal pipeline pressure. A pressure gauge for monitoring purposes is present and there may also be an alarm that operates if excessively high or low pressure develops. Pressure relief valves are present to prevent high pressure developing in the pipeline.

The main source of piped oxygen in large hospitals is not from cylinders but from a liquid oxygen supply. The underlying principle of the liquid oxygen supply is discussed in Chapter 10. This liquid oxygen supply usually has a capacity to last for at least six days but there is also a reserve manifold of cylinders. A control mechanism operates between the liquid oxygen supply and the bank of cylinders so that if the liquid supply fails for any reason this mechanism automatically turns on the cylinder supply.

ENTONOX SUPPLIES   Read

Oxygen is not the only gas which may be piped to theatres and intensive care areas from a central source. Entonox, nitrous oxide and air may also be supplied from a manifold of cylinders.

Entonox is a mixture of 50% oxygen and 50% nitrous oxide, stored in gaseous form in cylinders at a pressure of 137 bar. If a full cylinder of Entonox is cooled, a liquid phase, containing about 20% oxygen, 80% nitrous oxide, may form below the gas (Fig. 21.5).

Depending on the pressure in the cylinder, this may take place at a temperature as high as −5·5 °C. If the gaseous phase is drawn from a cylinder in which this condensation has occurred, the gas initially contains a high percentage of oxygen. However, the oxygen concentration falls as the oxygen passes out of the liquid phase to replace that drawn from the gaseous phase. Eventually, an anoxic mixture remains when most of the oxygen has passed out of the liquid phase. To prevent this occurring large cylinders have a dip tube which draws off any liquid phase first. This prevents the delivered oxygen concentration falling below 20% at any time. Nevertheless, this is only a secondary safety provision and it is recommended that the temperature in the storage area for these cylinders is maintained above 10 °C.

The nitrous oxide manifold is very similar to the oxygen manifold. In this case, the cylinders contain liquid nitrous oxide and so have a greater reserve of gas than the equivalent oxygen cylinders.

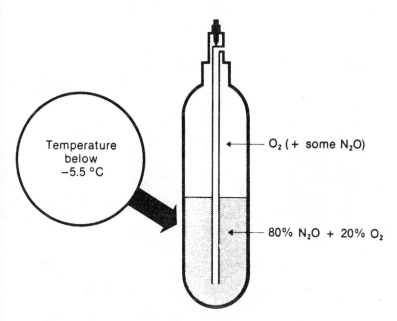

Figure 21.5   Effect of cooling below −5·5 °C on an Entonox cylinder.

COMPRESSED AIR SUPPLIES

Air is another supply often piped to theatres and although it can also be provided from a manifold of cylinders a central compressor plant is a more economical source for larger hospitals.

Figure 21.6 illustrates the main components of a central compressor plant. In practice, apart from the reservoir, the main units are duplicated so that any item can be serviced or repaired without interrupting the air supply. The air intake shown on the top left should normally be out-of-doors in such a position that it is not affected by rain, snow, dust or fumes. Air from the intake is drawn through a preliminary filter and a silencer into the compressor, which incorporates an after-cooler to cool the air once it has been compressed. The air then passes via a non-return valve into the large cylindrical receiver which acts as a reservoir to maintain a constant air pressure.

After leaving the reservoir the air is cleaned by means of separators and filters and then dried before it is piped to the theatres and wards. Most compressors are oil-lubricated, and consequently it is essential that oil mist is removed from the compressed air so that patients breathing it do not get oil pneumonitis. Coarser droplets of oil are baffled out in the separators, then finer particulate matter is removed in the filters. The air too must be dried because its relative humidity will have increased on compression. Two driers are present in each unit and

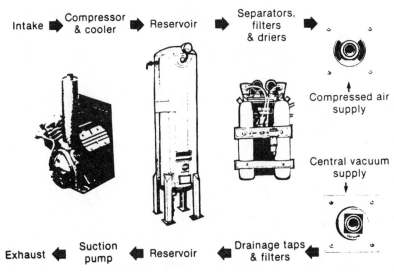

Figure 21.6 Compressed air supply and central vacuum supply systems.

they contain a suitable desiccant material—for example, silica gel. They are fully automatic and change automatically from the drier in use to the reserve should the air humidity rise. The humidity should be such that the dew point of the air supplied is less than $-40\,°C$.

A final filter follows the drier to ensure a bacteria-free air supply. The actual distribution pressure of the air is higher than for other piped gases; it is usually 6·9 bar because the piped air is commonly used to operate surgical tools. Alternatively, separate air supplies for surgical tools and for anaesthetic use may be provided.

CENTRALISED VACUUM SYSTEM

A centralised vacuum system is another supply often installed as part of a medical gas pipeline system and, as with the piped air supply, the vacuum system consists of a pump, a large receiver and a filter unit as shown in Fig. 21.6. In the bottom right of the figure is the typical vacuum terminal unit through which the air from wards and theatres is first drawn into drainage traps and bacterial filters to dry and clean it. The air then passes into a cylindrical receiver which acts as a reservoir to maintain a constant suction and also prevents the duty pump having to run continuously when the load is light. One pump is illustrated but there are normally two, each of which is capable of maintaining a vacuum of not less than 0·53 bar (400mmHg) below the standard atmospheric pressure of 1·01 bar (760 mmHg). The exhaust gases from the pump pass through a silencer before being discharged out-of-doors, usually away from windows or other air intakes. In some hospitals an active scavenging system for anaesthetic gases may be used that works on the same principle as the vacuum supply but usually without a vacuum reservoir.

SUCTION APPARATUS

Suction apparatus is connected to the vacuum pipeline; Fig. 21.7 illustrates a typical suction control. Application of a vacuum through the inlet evacuates the compartment below the diaphragm. The setting of the control knob at the top alters the pressure of the spring on the diaphragm. This in turn varies the pressure at which the diaphragm control valve opens or closes the inlet and so adjusts the degree of vacuum. The vacuum is indicated by a gauge. Below the control valve is a compartment with a float control and a filter and a collection jar with its own float control lies between this compartment and the patient. The two float controls protect the suction control system and the pipeline by occluding the airflow if the compartments become full and

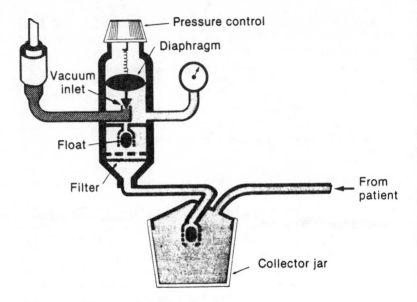

Figure 21.7   Suction control system.

the filter provides additional protection from particulate and nebulised matter.

Suction units normally provide a maximum flow of over 40 litre min^{-1} at a vacuum of 0·53 bar (53 kPa).

THE OXYGEN CONCENTRATOR

The oxygen concentrator has some similarities to the central compressor plant used to provide compressed air, but in addition incorporates cylinders containing zeolite. This is a silicate compound which has ion exchange properties and acts as a molecular sieve, separating the oxygen from nitrogen in the air. It is similar to materials used to pack some chromatographic columns.

Figure 21.8 illustrates the principle of its action. As in compressed air supplies the air is filtered, then compressed and cooled before being dried by silica gel. However, in addition to the silica gel the two cylinders shown also contain the zeolite which absorbs nitrogen. Solenoid valves direct the flow of gas so that one cylinder absorbs nitrogen and water vapour while absorbed gas from the other is drawn off by the vacuum pump. The solenoid valve reverses every half minute to give a steady flow of oxygen-enriched air to the reservoir.

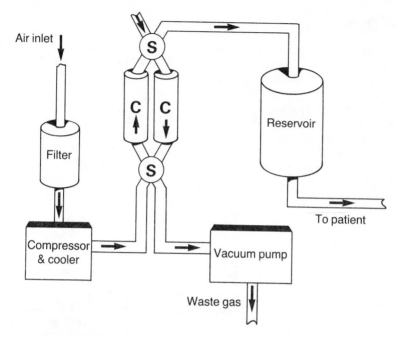

Figure 21.8   Oxygen concentrator. C = cylinder containing zeolite. S = solenoid valve.

Small oxygen concentrators are available for use in the home for long-term domiciliary oxygen therapy. They can give oxygen concentrations of 90 to 95% at a flow of 4 litre min^{-1} and are more convenient than the use of oxygen cylinders. The running cost of the electricity used and servicing is much less than that of cylinder or liquid oxygen supplies. Although 2 to 5% argon is present in the oxygen mixture produced by concentrators it appears to have no toxic effect. The balance of the gas delivered is mainly unabsorbed nitrogen and carbon dioxide.

A disadvantage of these concentrators is that the supply is restricted to a flow of up to 4 litre min^{-1}, and to a pressure of under 70 kPa. This is much lower than the 410 kPa provided by cylinder reducing valves, and the pipeline supplies in hospitals. Therefore, equipment using venturis and ventilators and designed for hospitals is unsuited to use with the smaller oxygen concentrators. However, larger oxygen concentrators are made for small hospitals and clinics in remote areas lacking reliable oxygen supplies, and these provide the gas at a higher pressure and flow.

SCHRAEDER VALVES

All the piped supplies to theatres and other areas finish in special terminal units with a non-interchangeable coupling; in the UK this is normally a Schraeder valve, illustrated in Fig. 21.9. The non-interchangeability is achieved by the use of varying sizes of collars on the Schraeder probe so that a given collar only fits the appropriate socket; thus the nitrous oxide collar is larger than the oxygen collar as shown. The Schraeder outlet is also labelled and colour coded and contains an internal non-return valve which seals the gas supply until the probe is plugged in. On unplugging, the valve closes off the gas supply.

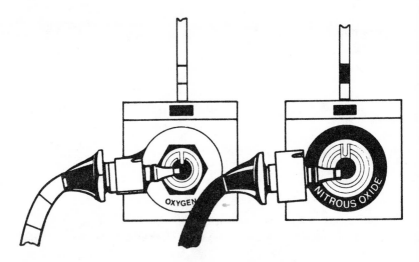

Figure 21.9 Schraeder valve system for medical gases.

LEAKS AND FAULTY CONNECTIONS

A leak at any point in the gas supply system or at the anaesthetic machine is a matter for concern. If a leak occurs at the pin-index mounting point it could indicate a fault at the non-return valve, which is usually incorporated between the cylinder and the reducing valve. Alternatively, it could be due to damage to a 'Bodok seal', the compressible washer present at this point.

Leaks at other points in the anaesthetic machine may result in the patient receiving less gas than intended and may also cause an increase in air pollution in the theatre.

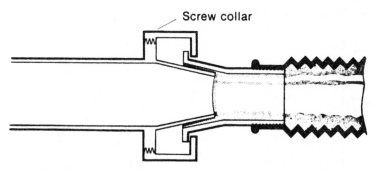

Figure 21.10   Screw collar system for use with anaesthetic tubings.

Leaks can also occur if connections in the anaesthetic tubing systems are not secure; connections are available with a screw collar to provide a more secure lock as shown in Fig. 21.10.

Great care is needed not only to avoid leaks but also to prevent any risk that anaesthetic system tubings or gas supply leads are connected incorrectly.

Suppose, for example, that unauthorised personnel have wrongly connected the nitrous oxide and oxygen tubings on the anaesthetic machine as shown in Fig. 21.11. The anaesthetist may not notice this

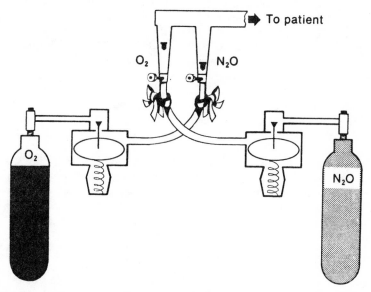

Figure 21.11   Risk of reversal of oxygen and nitrous oxide supply leads.

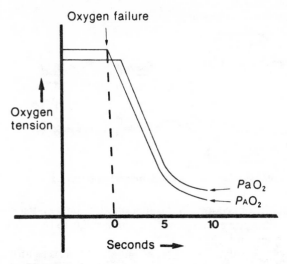

Figure 21.12   Fall of alveolar oxygen tension during ventilation with an anoxic gas.

immediately if he is using mixtures of oxygen and nitrous oxide, but it may result in the patient being awake during anaesthesia and in an emergency the patient may receive 100% nitrous oxide instead of 100% oxygen with fatal consequences.

Figure 21.12 illustrates the rapidity with which the alveolar oxygen tension falls if a pure anoxic gas such as nitrous oxide is administered to the patient. In five to ten seconds a dangerously low alveolar oxygen tension ($P_{A}O_2$) is achieved, followed almost immediately by a drop in the arterial oxygen tension (PaO_2). This fall is much more rapid than in the case of respiratory obstruction because when breathing is obstructed there is a reserve of oxygen in the lungs, whereas in the case shown in Fig. 21.12 the oxygen is being actively washed out of the patient's lungs and blood.

This problem may also occur if the oxygen supply fails or is allowed to run out while the nitrous oxide supply continues; most modern anaesthetic machines now incorporate safety warning devices and a nitrous oxide supply cut-out to protect the patient in the event of oxygen supply failure (Chapter 1).

REPAIRS TO PIPED GAS SUPPLY SYSTEMS

In addition to accidents caused by errors in gas supply systems at the anaesthetic machine, problems have arisen when repairs have been made to the piped gas supplies.

If maintenance or repairs are carried out on a piped medical gas system, it is important that the terminal units affected are not in use at the time. There are also several tests which must be carried out at every outlet before the system is put back into service.

Medical air is used for the first two tests. For the leak test the whole system is pressurised and sealed. Over a period of 24 hours there must be no unacceptable drop in pressure. An anti-confusion test is performed by connecting air to one pipeline system at a time, and making sure that the test air is delivered from every terminal unit bearing the name of the gas pipeline system being tested and that it is not delivered from any other terminal unit.

Once the pipeline systems have been connected up to their appropriate gas sources, gas identification tests must be carried out by trained personnel to check the identity and purity of the gases delivered at the terminal units. An oxygen-content analyser is used to check the identity, giving readings of 0% for nitrous oxide, 21% for air, 50% for Entonox and 100% for oxygen. A chemical gas detection tube may be used to test the purity of the gas delivered, detecting any trace level impurities which might have been introduced into the system by pipe-jointing procedures. Air from a compressed air supply should be tested further to ensure that it is free from impurities such as oil, carbon monoxide or carbon dioxide.

These procedures are primarily the responsibility of the hospital authorities but anaesthetists should have a general knowledge of them so that appropriate precautions are observed.

22

Breathing and Scavenging Systems

This chapter deals first with the systems by which gases or anaesthetic vapours may be delivered to the patient's airway.

THE OPEN SYSTEM

This is the simplest system and is illustrated in Fig. 22.1. In this system the patient's airway remains open to room air and no tubing, valves or reservoir bag are used. In the first example A, the anaesthetist is inducing a child by allowing the anaesthetic gases to fall over the child's face from a cupped hand containing the end of the gas delivery tube. In earlier forms of anaesthesia, a Schimmelbush mask was used, held over the patient's face in a similar manner. In example B, oxygen is delivered to an Edinburgh mask which has a wide opening to room air and gives a negligible increase to the respiratory dead space. In example C, oxygen is delivered by cannulae to the patient's nostrils. A similar technique has also been used during anaesthesia, a constant flow of anaesthetic gases being delivered by the side tube of a special gag into the mouth of the patient.

Open systems have the advantage of presenting no added resistance to the patient's breathing, but the disadvantage of giving little control over the concentration of oxygen and other gases received by the patient because air dilution is inevitable. During peak inspiration, a large inspiratory flow of up to 50 litre min^{-1} may occur, so dilution with air is high during this part of the respiratory cycle.

Known concentrations of gases can only be administered if the flow of gas to the patient exceeds his peak inspiratory flow, and some oxygen therapy 'venturi' masks achieve this by use of an injector technique (Chapter 2).

To administer more accurately controlled concentrations of gases a tightly fitting face mask or intubation is used and the main breathing systems may be categorised as non-rebreathing valve, T-piece, Magill

A   B   C

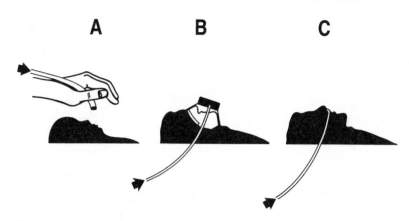

Figure 22.1   Open breathing systems.

and closed systems. All the systems have a reservoir in the form of tubing or a bag to allow adequate gas supply to the patient during peak inspiratory flow.

NON-REBREATHING VALVE SYSTEMS

A typical non-rebreathing valve is the 'Ambu E valve' illustrated in Fig. 22.2A. It contains moulded rubber one-way inspiratory and expiratory valves positioned to allow the patient to draw in gas through one connection and expire through the other. In the example shown, a reservoir bag is present in the system and the gas supply to this must equal the tidal ventilation. An alternative non-rebreathing valve system is found in the Quantiflex machines used in dental anaesthesia in which the inspiratory valve is positioned immediately after the reservoir bag.

The Ambu E valve can also be used for controlled ventilation as shown in Fig. 22.2B. Squeezing the reservoir bag applies positive pressure to the inspiratory port, moving the valve flap to occlude the outlet as shown. Thus, the Ambu E valve can act both as an inflating valve and as a non-rebreathing valve. In a variety of the valve, the Ambu E2, used only in resuscitation apparatus, the expiratory flap valve is omitted and the non-rebreathing function is then no longer present. It is important that these two types of valve are not confused, particularly when they are re-assembled after cleaning. Valves should be closely supervised in use lest an excessive flow of fresh gas maintains the valve in the inflation position (Fig. 22.2B), giving a dangerous build-up of airway pressure.

Because the pressures needed to overcome the inspiratory and

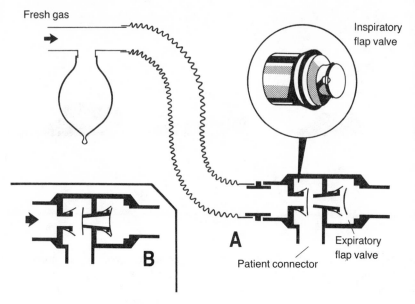

Figure 22.2 Non-rebreathing system with Ambu E valve. (A) During expiration. (B) During positive pressure inspiration.

expiratory valve resistances can be 100–200 Pa (1–2 cmH₂O) during peak flow, many non-rebreathing valves are uncomfortable for long-term use in conscious individuals. In breathing apparatus used for airmen, firemen and divers, non-rebreathing systems have special low resistance flap valves and the Entonox apparatus used for analgesia has a similar low resistance expiratory valve . In place of the inspiratory valve a special demand valve is used for Entonox as described in Chapter 1, and similar demand valves are used to deliver air or gaseous mixtures in breathing apparatus.

A non-rebreathing valve system is useful for emergencies outside hospital as it allows anaesthesia to be administered even when cylinder gas supplies are not available. Figure 22.3A illustrates one such system, the Triservice apparatus. Inspiratory and expiratory valves are incorporated in the Laerdal valve and, as in the Ambu E, the inspiratory valve can also function as an inflating valve (Fig. 22.3B). Thus, the silicone rubber flaps of the valve occlude the expiratory ports when the self-inflating bag is compressed. Air is drawn through a Triservice vaporiser which is a modified OMV vaporiser (Chapter 11), and the T-junction shown allows oxygen supplementation should supplies be available. The advantages of this system are its extreme lightness and

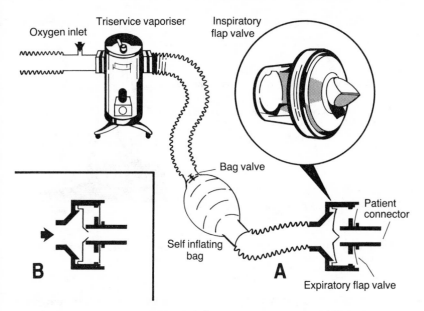

Figure 22.3 Triservice apparatus. (A) During expiration. (B) During positive pressure inspiration.

portability. Similar systems are also possible with other draw-over vaporisers.

T-PIECE SYSTEMS

The T-piece system is sometimes known as Ayre's T-piece after Dr Philip Ayre who first used it in infants in 1937. There are no valves in the simplest form of this system, fresh gas being supplied to a T-piece at the patient who breathes through an open limb of the T-system as shown in Fig. 22.4. This limb must be of suitable diameter to avoid excessive resistance to the flow of gas, e.g. 22 mm internal diameter for an adult patient.

In a typical system a fresh gas flow of 2 to 2·5 times the tidal ventilation (e.g. 14 litre min^{-1}) is supplied to the T-piece and the capacity of the open limb could be the same as the patient's tidal volume. The figure illustrates the content of the open limb at the end of expiration, when it contains a mixture of expired and fresh gas. The gases from the patient's anatomical dead space are expired first, so these are shown released from the open end of the tube at A. During mid-

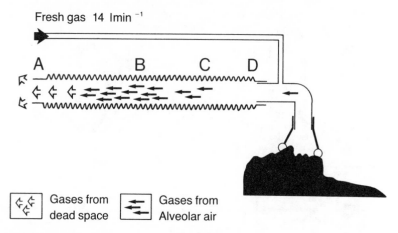

Figure 22.4   T-piece system. Gas distribution towards the end of expiration (see text).

expiration, the majority of the alveolar gases are expired at near peak expiratory flow, so they are only slightly diluted by the fresh gas flow as shown in the centre of the tube at B. At the finish of expiration the expiratory flow falls, so that the end of the tube nearest the patient, C, contains more fresh gas. If there is an end-expiratory pause, then a further volume of undiluted fresh gas is finally present as at D in the figure.

Consider now what happens during inspiration. Initially, fresh gas only is inspired until inspiratory flow becomes greater than the flow of fresh gas to the T-piece. At peak inspiration, the patient may in addition draw upon the gas at D in the open tube. Provided there is an end-expiratory pause this is fresh gas, and so no rebreathing of alveolar air occurs.

If the fresh gas flow is too low, e.g. below twice the tidal ventilation, the patient will draw upon the gases at B and C in the open limb during peak inspiration and rebreathe alveolar gases.

Figure 22.5 illustrates the effect of the end-expiratory pause upon the fresh gas flow required to prevent any rebreathing of alveolar gas. The graph shows the gas volume inspired from the open limb of a T-piece by an anaesthetised patient with a tidal volume of 0·4 litre, breathing at 12 per minute. The respiratory cycle therefore lasts 5 seconds.

The difference in the vertical height between the inspired gas volume and the fresh gas volume supplied is the volume of fresh gas in the open limb. If at any time the volume of inspired gas exceeds that of fresh gas supplied and stored in the open limb, rebreathing will occur. It can be seen from the graph that a flow of 14 litre min^{-1} (2·9 times the given

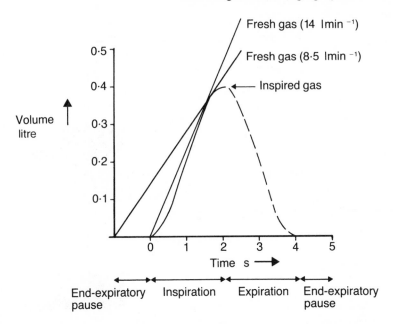

Figure 22.5 T-piece system. Effect of an end-expiratory pause upon the fresh gas flow required to prevent rebreathing of expired alveolar air (see text).

tidal ventilation) is necessary if there is no end-expiratory pause. Provided that an end-expiratory pause of 1 second is present, then a flow of fresh gas of 8·5 litre min^{-1} (just under twice the tidal ventilation of 4·8 litre min^{-1}) ensures that the volume of fresh gas supplied always exceeds the volume of inspired gas from the open limb.

At the T-piece, the patient inspires from the fresh gases as well as from the open limb, so this limb need not be as large as the full tidal volume. Nevertheless, if the open limb is omitted, this leads to inspiration of air unless the fresh gas flow is equal to the patient's peak inspiratory flow.

It is possible to add an open-ended reservoir bag (Fig. 22.6A), or a bag with a valve to the open end of the T-piece (Fig. 22.6B). Movements of the bag indicate respiration and the bag can be used to control ventilation. When the T-piece is used with controlled ventilation it is possible to use lower fresh gas flows, because tidal ventilation can be increased to compensate for the rebreathing of alveolar gas from the open limb. In this case too, the pattern of ventilation can be adjusted to give an adequate end-expiratory pause to render the system more effective.

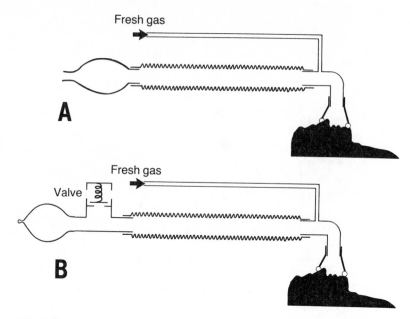

Figure 22.6 T-piece system. (A) With bag. (B) With bag and valve.

The T-piece system is particularly suited to use in paediatric anaesthesia because the T-piece connection can be close to the patient's airway with negligible apparatus dead space at this point. In addition, in its simplest form, it avoids the problems associated with an expiratory valve, such as increased resistance to flow and possible malfunction due to sticking. In adults, a disadvantage of the system is the high fresh gas flow which is required during spontaneous breathing. Apart from the cost the added flow increases expiratory resistance and may raise the risk of pollution from spilt gas.

COAXIAL T-PIECE SYSTEM

A coaxial T-piece system is a more compact form in which the gas delivery tube of the T is inside the limb of the T (Fig. 22.7). Typical examples are the Bain (USA) or Penlon coaxial (UK) systems. As in other T-piece techniques, gas flows during spontaneous breathing should be about 2·5 times the patient's tidal ventilation. Special care is needed in these systems to ensure that the central tube carrying fresh gas is correctly attached, otherwise the fresh gases may not be delivered to the patient end of the tube.

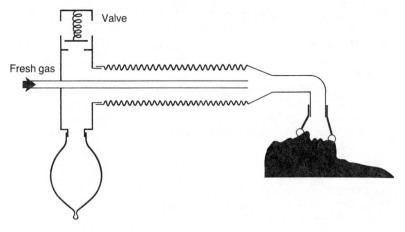

Figure 22.7   Coaxial T-piece system.

MAGILL SYSTEM

The Magill system (Fig. 22.8) was introduced by Sir Ivan Magill in
1920. It comprises a reservoir bag, delivery tube and expiratory valve
and remains a popular system for administration of anaesthesia in a
spontaneously breathing patient. The figure illustrates the contents of
the tube at the end of expiration. Gases from the anatomical dead space
travel along the tube at A, while fresh gas inflates the reservoir bag,

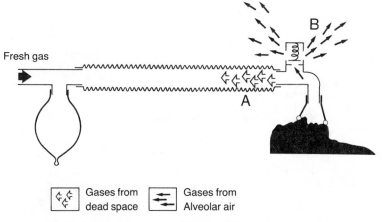

**Figure 22.8   Magill system. Gas distribution towards the end of
expiration.**

distending it until the pressure from the bag is sufficient to open the expiratory valve. This could be at 50 Pa or 0·5 cmH₂O pressure. Gases from the patient's alveolae are then expired through the expiratory valve into an expired gas scavenging system, or released into the atmosphere.

Because the Magill system can conserve the dead space gas it can be used at fresh gas flows of 70% of the patient's tidal ventilation in a spontaneously breathing patient.

The Magill system can also be used to ventilate the patient, the expiratory valve being tightened to a suitable extent and the reservoir bag squeezed to give intermittent positive pressure to the limits set by the valve. Excess gas spills from the valve at the end of inspiration. Because the gas spilt in this case is mainly fresh the system is inefficient during controlled ventilation, but may be convenient for short-term use while an alternative, e.g. a ventilator, is arranged.

Modern expiratory valves are larger than the former 'Heidbrink' type expiratory valves as they incorporate a connection for a scavenging tube. However, if the bulk of the valve is inconvenient then it can be positioned remote from the patient, as shown in Fig. 22.9A. A Magill system with remote valve can be convenient if it is decided to use disposable tubing to achieve better sterility, as the valve need not then be changed with each fresh tubing.

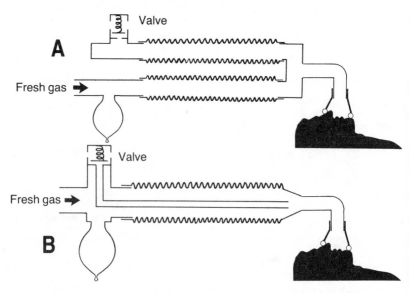

Figure 22.9(A) Magill system with remote expiratory valve. (B) Coaxial Magill system.

COAXIAL MAGILL SYSTEM

The coaxial Magill system also uses a remote expiratory valve, as illustrated in Fig. 22.9B. In the system shown, known as the Lack system, the fresh gas flows from the reservoir bag through the wide tube which contains a narrower tube leading to the expiratory valve. Unlike the coaxial T-piece the patient breathes through both tubes in the coaxial Magill system. Hence, wider coaxial tubes are required to prevent increased resistance to breathing. The same precautions must be taken as in other coaxial systems to ensure that the internal tube is correctly fitted and positioned. As in the case of the standard Magill system, the coaxial Magill is less suitable for prolonged controlled ventilation.

CLOSED SYSTEMS

In the fully closed system, also known as the closed circuit or circle system, the patient's gases recirculate through soda lime which absorbs carbon dioxide, while oxygen is added to replace the patient's metabolic uptake. A typical arrangement is illustrated in Fig. 22.10, but the positions of the components may vary.

The soda lime used must have granules of a critical size, large enough to avoid excess resistance to breathing and small enough to absorb carbon dioxide efficiently; a typical size being '4–8 mesh', indicating that the granules will pass through a sieve having four strands per inch, but not through one with eight strands per inch.

When the system is used for anaesthesia, anaesthetic vapour is added in the concentration needed to meet clinical requirements and the uptake by the patient. The anaesthetic may be supplied from a draw-over vaporiser in the circle (VIC), or by an out of circle vaporiser (VOC). The latter needs to be a special type which retains its accuracy at low flows. Because of the variability of the patient's uptake of oxygen and anaesthetic agents, gas analysers are recommended when the closed system is used. This is to ensure that the concentrations of oxygen, carbon dioxide and anaesthetic vapours are satisfactory.

Many anaesthetists find the cost of these analysers, and their maintenance, limits the use of the completely closed system. Other disadvantages are the difficulty of sterilisation because of the size and complexity of the closed system compared with the simpler versions, and the need for a special form of ventilator. The closed system has the advantage that it allows the oxygen uptake at the patient to be monitored, and the uptake of volatile anaesthetic agents to be demonstrated. Furthermore, the heat loss from the patient is slightly reduced because the inspired gas is fully humidified at near ambient

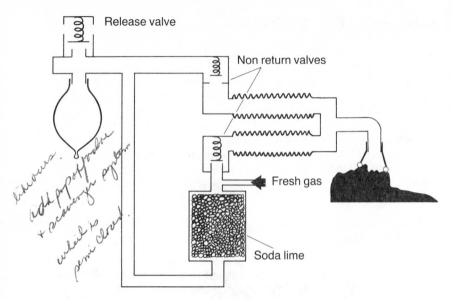

Figure 22.10 Circle system. When fully closed, the fresh gas flow is oxygen at a rate equivalent to the patient uptake, with additional anaesthetic vapour as required. The reservoir bag illustrated can be replaced by a calibrated spirometer to indicate volume changes.

temperature (Chapter 10). The economy in the amount of volatile anaesthetic used must be balanced against the added cost of the soda lime required. Another advantage, the reduction of pollution, is less relevant now that scavenging systems are recommended.

If a fresh gas flow of, for example, 2 litre min^{-1}, is supplied to the closed system and excess gas spilt from the expiratory valve, the concentrations of gases in the system approximate to those in the fresh gas supply. This system, sometimes described as a closed system with leak, gives better control of the concentrations but lacks some of the advantages claimed for the fully closed system.

If nitrous oxide were to be used with a fully closed system, the presence of nitrogen in the system would prevent the build-up of an adequate concentration of nitrous oxide. The use of a closed system with leak, at least for the initial period of use, allows the nitrogen to be replaced by nitrous oxide.

ALTERNATIVE CLASSIFICATION OF SYSTEMS

Unfortunately, there is no international standard for the classification of breathing systems. Even the word system is not universal as some use the word 'circuit' instead. Nevertheless, the term 'system' is preferred in the classification given here, as only in the closed system does the breathed gas complete a circuit.

In the UK, systems such as the Magill, which are neither closed nor open are often referred to as 'semi-closed', but in the USA the term 'semi-open' is used instead; the term 'semi-closed' being used for the closed system with leak. In view of this confusion the terms semi-open and semi-closed are best avoided.

Following an analysis of the performance of some breathing systems by Mapleson, the letters used by him to identify systems are as follows;

	Mapleson
T-piece system (unmodified)	E
T-piece system with open ended bag	F
T-piece system with bag and valve	D —
Magill system	A

(The systems B and C in this classification are now seldom used and not described in this chapter. The open, non-rebreathing valve, and closed systems are not included in this categorisation.)

RESISTANCE OF BREATHING SYSTEMS

The open system gives no added resistance to breathing but in other systems resistance occurs in the tubings and valves used.

If the tubing diameter is large enough to allow laminar flow then the additional pressure needed to overcome the resistance during inspiration and expiration is very small. As explained in Chapter 2 the critical flow for a smooth tube of 22 mm internal diameter carrying a nitrous oxide–oxygen mixture could be about 22 litre min^{-1}. Anaesthetic systems use corrugated tubing and peak flows may exceed 22 litre min^{-1}, so both laminar and turbulent flow usually occur during the breathing cycle. Nevertheless, the extra pressure the patient must generate to drive the gases through the tubing is small, perhaps about 25 Pa (0·25 cmH$_2$O) during peak flow.

In patients having respiratory investigations involving the breathing of air or oxygen mixtures, even this small pressure may be undesirable, so tubing of a larger diameter, e.g. 30 mm, is used. Because laminar flow resistance is inversely proportional to the fourth power of the radius, the increase of diameter from 22 mm to 30 mm gives a 3·5 times reduction in resistance to laminar flow. In addition, the increased diameter reduces the risk of turbulent air flow which would give a sharp rise of resistance.

In simple T-piece systems without any expiratory valve or reservoir bag the resistance is that of the limb of the T through which the patient breathes. In practice, the pressure required to be generated by the patient to overcome this resistance is greater during expiration than during inspiration. This arises because expiratory flow is increased by the fresh gas flow, whereas inspiratory flow in the limb is reduced by the fresh gas flow. The higher the fresh gas flow the greater will be the back pressure during expiration. This factor applies generally in most breathing systems.

The valves are an important source of resistance in other breathing systems. A typical valve may have an opening pressure of 50 Pa (Chapter 1). However, during peak flows of about 30 litre min^{-1} the turbulence in the valve results in higher pressures of 100–500 Pa (1–5 cmH_2O), depending upon the valve concerned.

The reservoir bag used in a system can also affect the pressures which develop during spontaneous breathing.

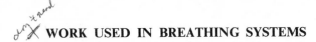

WORK USED IN BREATHING SYSTEMS

The additional work required in breathing systems can be indicated by the area of a pressure–volume loop (Chapter 8). Figure 22.11 illustrates the type of loop for various systems. The pressure at the patient connector is plotted against the volume change for a tidal volume of 0·4 litre.

As explained above, the main source of resistance is in the valves of a system. Therefore, the simple T-piece system gives very little added work, as shown by the small area of the loop in Fig. 22.11A. In the case of the non-rebreathing valve and circle system (Fig. 22.11B) or the Magill system (Fig. 22.11C) the mean pressure swing could be 200 Pa (2 cmH_2O) for the volume change of 0·4 litre. The area of the loop and the work done during one ventilatory cycle would then be as follows:

$$\text{Additional work done} = 200 \text{ Pa} \times (0.4 \times 10^{-3}) \text{ m}^3$$
$$= 0.08 \text{ J}$$
$$= 80 \text{ mJ}$$

The total mechanical work for a normal inspiration could be 300 mJ, so a typical breathing system could give a 27% increase in the work of breathing. Such increases of work are uncomfortable for conscious individuals but, as already explained, special valves with lower resistances are then used.

In the anaesthetic non-rebreathing valve system the resistance of the inspiratory and expiratory valves are comparable, so the pressure–volume loop illustrated (Fig. 22.11B) shows that the inspiratory and expiratory work are similar as indicated by the equal areas above and below the zero pressure axis. A circle system would give a similar result.

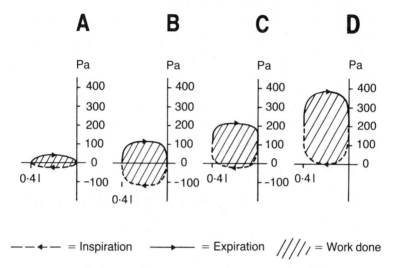

Figure 22.11 Work in breathing systems. Graphs of the pressure change at the patient's airway for a tidal volume of 0·4 litre (spontaneous ventilation). **(A)** Simple T-piece system. **(B)** Non-rebreathing valve or circle system. **(C)** Magill system. **(D)** T-piece with reservoir bag and valve.

In the Magill system there is no inspiratory valve and the assistance to inspiration from the slightly distended reservoir bag and from the fresh gas flow overcomes the small flow resistance of the 22 mm diameter tube. Consequently, all the work required from the patient to overcome the system resistance occurs during expiration, and it is seen (Fig. 22.11C) that the pressure–volume loop is almost entirely above the zero pressure axis. The coaxial Magill system performs similarly, but the expiratory resistance in this case is that of the valve with its central 12·5 mm diameter supply tube.

The T-piece system with a bag and valve could give a similar tracing if gas flow were comparable, but a high fresh gas flow of 2 to 2·5 times tidal ventilation is recommended. In consequence the mean expiratory flow through the valve and tubing is increased, and much higher pressures are generated in expiration as shown in Fig. 22.11D. The high gas flow maintains a positive pressure during part of an inspiration assisting this phase of the cycle, and so the patient uses less energy during inspiration. A similar tracing could be obtained with the coaxial T system.

In anaesthesia in a spontaneously breathing patient, if the increase in the work of breathing required with a system is excessive, it may augment the depression of ventilation which occurs. The added expiratory resistance gives a slight rise in intrathoracic pressure and so could reduce venous return and cardiac output.

Because a face mask rarely gives a completely airtight fit, a system which raises the mean inspiratory pressure inside the mask has the advantage that air dilution of the anaesthetic gases during inspiration is reduced. During expiration, on the other hand, the leakage of the gases round the mask increases if the mask pressure is raised, and so air pollution is increased.

SAFETY OF BREATHING SYSTEMS

The components of systems must be correctly assembled and the connections of systems are designed so that there is normally a progression of male to female cone fittings in the direction of the flow. There should be no risk of the connections falling apart accidentally, and screw collars are available (Chapter 21). However, final connection at the patient is normally a simple cone fitting to allow rapid change of this connection as needed. Plastic connections may distort during autoclaving and even metal ones can be damaged by misuse, so regular inspection is recommended with replacement as necessary.

Breathing tubes should not be too narrow and sharp angles at connections should be avoided as they not only increase the incidence of turbulent flow but also the resistance to breathing (Chapter 2).

Grossly excessive pressures can arise if the expiratory valve sticks or if it is omitted from the breathing system. Alternatively, excess pressure can build up if the scavenging tubing is obstructed. In such cases the pressure builds up to the limit governed by the reservoir bag, if one is present. This pressure depends on the tension T in the wall of the bag and on the radius R, according to Laplace's law:

$$P = \frac{2T}{R}$$

When the bag is distended, the tension in its wall due to its elasticity is proportional to the circumference of the bag, and this in turn is directly proportional to the radius. Hence, the bag gives a pressure that is nearly constant, even though it becomes grossly distended.

This pressure is usually about 4 kPa (40 cmH$_2$O) and this gives some short-term protection to the patient while the anaesthetist identifies and removes the obstruction in the system.

Some disposable reservoir bags lack elasticity and give no such protection so that pressures in the system then rise to much higher values. If the system lacks a reservoir bag entirely then the pressure may rise to the limit set by the safety valve on the anaesthetic machine.

Typically this could be 35 kPa and such a valve only protects the mechanical components of the machine and gives no protection to the patient.

In connecting any breathing system, the T-connection, or in the standard form of Magill system the expiratory valve, must be positioned close to the patient's airway so there is no excess anatomical dead space. An example of an accidental increase of dead space arises if the central tube of a coaxial system becomes detached. The entire coaxial tube may then be added to the dead space and the patient may suffer hypoxia and hypercapnia in consequence.

VENTILATORS

A full description of ventilators is beyond the scope of this book. However, as an example, Fig. 22.12 shows in simplified form the principle of a ventilator of the Manley Pulmovent type.

A fresh gas flow of, for example, 7 litre min^{-1} inflates a reservoir bellows. The gas in the bellows is at a pressure of about 10–12 kPa due to the force exerted on the bellows by springs. When the bellows are

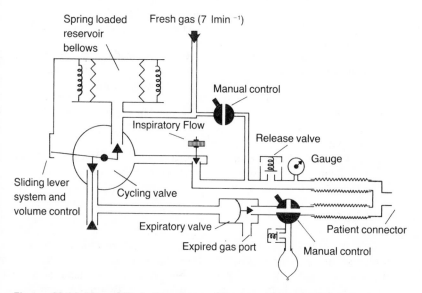

Figure 22.12 Simplified diagram to illustrate the principle of a ventilator of the Manley Pulmovent type. The ventilator is shown at the start of inspiration with both control ports of the cycling valve open.

inflated to the preset volume, the cycling mechanism opens a valve to allow the bellows to inflate the patient's lungs. The full pressure of 10 kPa is not however applied, as a flow control valve restricts the flow as shown and a safety valve at the patient's side of the cycling valve limits pressure to a maximum of 7 kPa. The cycling valve also controls the expiratory valve, so that it is closed during inspiration and open at expiration. During expiration the airway is open to atmosphere through the expired gas port while the reservoir bellows re-inflates.

It is convenient to classify ventilators as 'constant pressure generators', in which the gas is supplied at a constant pressure during inspiration, and 'constant flow generators' in which gas is supplied at constant flow. However, the Manley exemplifies a general problem in this classification in that, although a near constant pressure is generated it is applied via a flow control, so that neither pressure nor flow is constant at the patient's airway.

The cycling valve is tripped through levers by the tidal volume in the bellows, a control allowing the volume to be adjusted to meet requirements, so it might be regarded as a 'volume cycled' ventilator. However, clear-cut classification of the cycling of ventilators is also fraught with difficulty as the volume from the bellows is supplemented by the fresh gas flow in this case. The total minute volume is the fresh gas flow delivered from the anaesthetic machine. This is 7 litre min^{-1} in the example so, if the tidal volume measured by a Wright respirometer is 0·5 litre, the respiration rate is 14. Thus, the ventilator is sometimes referred to as a minute volume divider.

Figure 22.12 also shows the two controls needed to convert the ventilator to a system for manual control or spontaneous breathing. One control bypasses the reservoir bellows and cycling valve so that the fresh gases are delivered directly to the patient, while the other control bypasses the expiratory control valve, replacing it with a reservoir bag and expiratory valve. The breathing system is thus converted to a T-piece system.

Some forms of Manley Pulmovent also incorporate an extra bellows to give a facility for negative pressure during expiration.

SCAVENGING

There is an increased incidence of spontaneous abortion in female staff employed in operating theatres where anaesthetics are used and, although it is difficult to implicate conclusively the toxicity of trace quantities of anaesthetic agents in room air, it is reasonable to reduce the level of pollution by scavenging the expired anaesthetic gas.

Systems can be classified as active, in which an external source of power such as a pump draws away the scavenged gas, or passive in which the gases are driven to the exterior of the building by the pressure generated by the patient during expiration.

The techniques for active scavenging can be divided into systems resembling certain breathing systems, namely an open system, T-piece system, and one with reservoir bag and valves.

OPEN SCAVENGING SYSTEMS

In this type the point at which suction is applied to the waste gases is open to the atmosphere without any intervening reservoir bag or valves. The technique is generally less effective than the other systems. Techniques have been tried in which a funnel is positioned near the expiratory valve and mask (Fig. 22.13A) to aspirate spilt anaesthetic agents, e.g. in anaesthesia for dental extractions in which an airtight mask fit is difficult to achieve. Alternatively, the funnel can be inverted in the form of a dish and used to collect gases released from the end of a T-piece breathing system. In a patient in the recovery area or after Entonox analgesia the gases released from the airway may be scavenged by a funnel positioned close to the mouth (Fig. 22.13B). Like the open breathing systems, the open scavenging systems lack control. A high scavenging flow is needed and the funnel must be very close to the point of release of the gases to be effective.

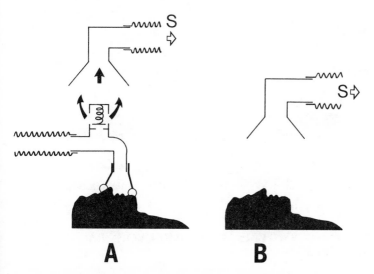

A   **B**

Figure 22.13 Open scavenging systems. Scavenging flow is indicated at S.

T-PIECE SCAVENGING SYSTEM

In this system the point at which suction is applied is in the form of a T-piece, one limb of which is open to the atmosphere. To obtain better efficiency the open limb may be expanded in the form of a reservoir (Fig. 22.14A). It is essential that the open end of the T-limb or reservoir is designed so that there is no risk of obstruction, which could result in either excessive subatmospheric or excessive positive pressure being applied to the patient's airway. The system is designed so that the scavenging flow is sufficiently great to remove all expired gas. A flow control and flow indicator are usually provided and the reservoir may incorporate a whistle activated by the patient's breathing. The scavenging flow required depends upon the size of the open T-tube or reservoir. Thus, if flow is equal to the peak flow in expiration a very short open limb, as in Fig. 22.14B, suffices.

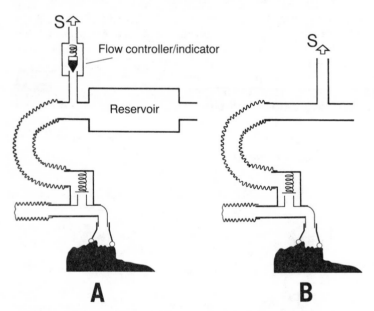

Figure 22.14 T-piece scavenging systems. Scavenging flow is indicated at S.

SCAVENGING SYSTEMS WITH RESERVOIR BAG AND VALVES

In this system a reservoir bag intervenes at the point at which suction is applied, and valves are present to prevent excess pressures (Fig. 22.15).

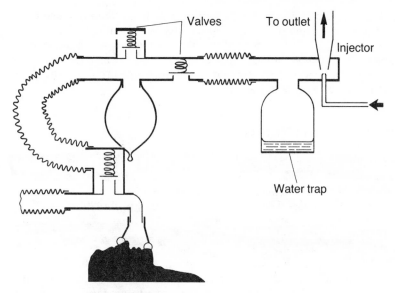

Figure 22.15   Scavenging system with bag and valves. The system becomes a passive one if it is used without the injector illustrated.

Two valves are illustrated in the figure, one to spill gases should excessive pressure build up and the other to protect against excessive subatmospheric pressure. These are relief valves and act as safety devices, unlike those in the breathing systems which are to direct gas flow in the system. One form of the reservoir bag and valve system used in the UK is known as the Papworth safety block.

In the system shown an injector is used to remove the expired gas. The water trap on the right collects condensed water.

SOURCE OF SCAVENGING FLOW

Some hospitals use the same system as the central vacuum type described in Chapter 21, but an independent system is usually recommended because it was thought that anaesthetic vapours could affect lubrication oils in the standard suction pumps. For individual use in areas where there is only a single anaesthetic machine, it is possible to use a separate pump or an injector powered by the compressed air supply (Fig. 22.15). Oxygen has been used in place of compressed air but gives an increased fire risk. A system using an injector gives a lesser vacuum than hospital suction. The use of low vacuum does not avoid the need for precautions at the point of collection at the patient, as even low pressure can give a risk of airway obstruction and damage or of pulmonary oedema if it is applied directly to the patient's lungs.

For safety, it is recommended that the positive pressure at the patient during scavenging should not exceed 50 kPa (0·5 cmH$_2$O) at a flow through the system of 30 litre min^{-1}, and that pressure should never be more than 50 kPa below atmospheric.

PASSIVE SCAVENGING SYSTEM

In the passive system a wide bore tube conducts the expired gases outside the building. This technique can give problems, as excessive positive or subatmospheric pressures may be caused by wind movements at the outlet. Furthermore, if the outlet is above roof level, as recommended to prevent re-entry of the scavenged gases to the building, then the weight of denser anaesthetic vapour such as nitrous oxide gives a back pressure in the patient's breathing system.

The collection system with bag and valves shown in Fig. 22.15 has been used without the injector as a basis for a passive scavenging system. To avoid the pressure problems due to air movement at the outlet of the passive system, the expired gases can be conducted to the exit grill used for the theatre ventilation. For this disposal system to work effectively it is essential that the theatre ventilation exit flow is suitable, as flows in the ventilation ducts may vary and reverse flow is not unknown. The use of an exit grill is also inappropriate in theatres in which the ventilation recirculates the air.

THEATRE VENTILATION

Regardless of the scavenging system used, some spillage of anaesthetic vapours is inevitable, e.g. at the end of anaesthesia, at induction, or when using a face mask. Efficient theatre ventilation, e.g. 15 air changes per hour, is therefore recommended in areas where volatile anaesthetics are used.

23

Fires and Explosions

GENERAL PRINCIPLES

In a fire or explosion a combustible agent or fuel combines with oxygen or another oxidising agent to give reaction products and the release of energy (Fig. 23.1). Before the reaction can start, a small amount of energy known as the activation energy must be supplied. The energy produced in the reaction appears as heat energy, which increases the temperature of the mixture. Once the temperature exceeds a certain level the reaction becomes self-sustaining, as the energy produced is then sufficient to supply the activation energy to propagate the reaction.

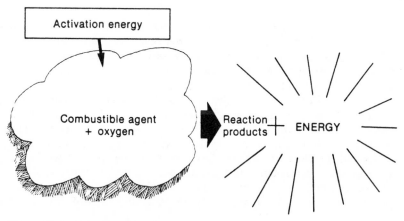

Figure 23.1   Principle of combustion.

Figure 23.2 shows the reaction for cyclopropane. The activation energy could be provided by a small spark. Two molecules of cyclopropane combine with nine molecules of oxygen to give six molecules each of carbon dioxide and water, and energy is released. If the initial mixture contains two volumes of cyclopropane to nine

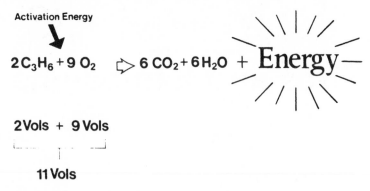

Figure 23.2   Combustion of cyclopropane.

volumes of oxygen, i.e. an 18% concentration of cyclopropane, all the cyclopropane and oxygen are used up and transformed into the product molecules, carbon dioxide and water. This concentration is known as the stoichiometric concentration.

> *The stoichiometric concentration* of any combustible vapour and oxidising agent is the concentration at which all the combustible vapour and oxidising agent are completely used up.

The most violent reactions take place in stoichiometric mixtures. If the proportions differ from this mixture, the reaction is less violent and when one component is greatly in excess of the other the mixture cannot be ignited at all. The limits outside which the mixture will not burn are known as the flammability limits.

Illustrated in Fig. 23.3, are the flammability limits for cylcopropane in oxygen and in air. The limits in oxygen are 2·5 and 63% and within these is the narrower range within which explosions may occur. The usual anaesthetic concentrations for cyclopropane are within these limits and therefore carry risks of explosion. As the speed at which the reaction spreads through a mixture depends on the proportions of fuel and oxygen present, a fire rather than an explosion is likely to occur at proportions nearer the flammability limits, as illustrated on the left of Fig. 23.4.

In such a reaction the pressure is close to atmospheric (1 bar) and the temperature is several hundred degrees Celsius. For proportions near the stoichiometric mixture in oxygen, on the other hand, the reaction may spread faster than the speed of sound and when this occurs an explosion rather than a fire takes place. When a mixture explodes, the reaction can spread at eight times the speed of sound and generate pressures as great as 25 bar and temperatures as high as 3000 °C.

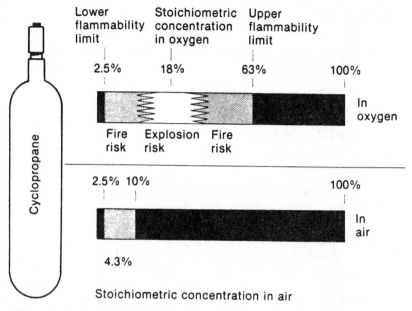

Figure 23.3 Flammability limits of cyclopropane in oxygen and in air.

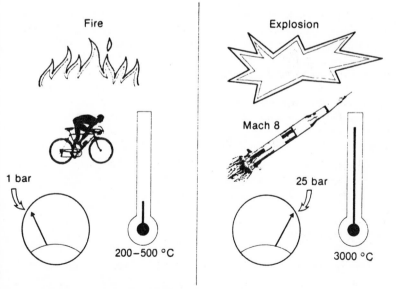

Figure 23.4 Comparison of conditions in a fire and an explosion.

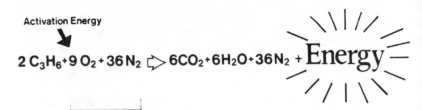

$$9(O_2 + 4N_2)$$

2 Volumes   **45** Volumes
Cyclopropane   Air

Figure 23.5   **Reaction in a stoichiometric mixture of cyclopropane and air.**

The violence of the reaction may be moderated by the presence of other molecules in a mixture.

As shown in Fig. 23.3 the flammability limits of cyclopropane in air differ from those in oxygen. In normal anaesthetic practice cyclopropane burns but does not explode in air; the range of flammability is reduced and the stoichiometric concentration is altered to around 4%. The inert nitrogen molecules not only absorb some of the energy produced but also do not take part in the reaction. Hence, the reaction is much less violent.

Figure 23.5 shows the reaction in the stoichiometric mixture in air. The ratio of cyclopropane to oxygen is the same in this case, but there are four nitrogen molecules present for every oxygen molecule, so the mixture contains two volumes of cyclopropane and 45 volumes of air to give a stoichiometric concentration of around 4%.

Nitrous oxide also is an oxidising agent. Nitrous oxide rapidly breaks down in the presence of the heat of a fire or explosion to release oxygen, giving a 33% oxygen mixture and releasing further energy:

$$2N_2O = 2N_2 + O_2 + \text{Energy}$$

An explosion can be more violent when nitrous oxide is present than with oxygen alone because, when a fuel burns in nitrous oxide, more heat is produced than when the fuel burns in oxygen.

ZONES OF RISK

The anaesthetic machine and the airway of the patient are the main areas of risk if flammable anaesthetics are used. In addition, gases may be spilt from the expiratory valve and from any leaks in the system;

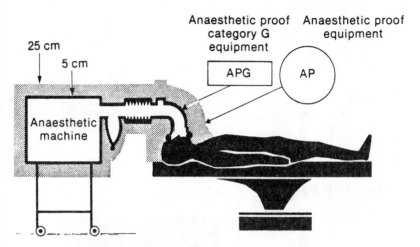

Figure 23.6   Zones of risk when flammable anaesthetics are in use.

hence there is a surrounding zone where the concentrations of gases may be flammable.

In the European Economic Community a zone of risk has been defined. This extends to 25 cm outwards from breathing systems and other parts, including the patient's respiratory tract, which contain flammable or explosive gas mixtures. There is also a zone of especially high risk extending to 5 cm (Fig. 23.6).

Equipment used within the zone of risk must be designed and constructed so that there is no possibility that it will ignite a flammable mixture—e.g. electric sparks must not occur. The effective zone of risk may be extended if anaesthetics are released under drapes, thus giving rise to locally high concentrations.

Equipment intended for use within the zone extending from 5 cm to 25 cm is known as anaesthetic proof equipment and, where possible, it is marked with the letters AP on a green circle.

Equipment intended for use within the zone of high risk extending to 5 cm is known as anaesthetic proof category G equipment and, where possible, it is marked with the letters APG on a green band.

OTHER FLAMMABLE AGENTS

Ether behaves in a similar fashion to cyclopropane, as illustrated in Fig. 23.7. At the top of the diagram are the flammability limits of ether in oxygen, 2 to 82%, and the stoichiometric concentration in oxygen (about 14%) at which there is a risk of explosion. Below are the

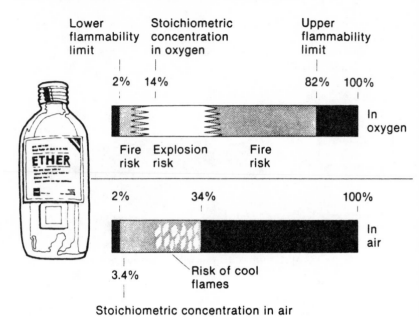

Figure 23.7   Flammability limits of ether in oxygen and in air.

flammability limits of ether in air. Like cyclopropane, ether does not explode in air but burns within the concentrations shown.

The concentrations of ether in oxygen used in anaesthesia are within the limits for an explosion risk and within the flammability limits for ether in air. A special risk may arise if liquid ether is spilt and produces a high concentration on the floor. The concentration of ether may approach the upper limits of flammability in air. If the ether were ignited, it could give rise to what are known as 'cool flames' which are invisible, and, as the name implies, are at a much lower temperature than normal flames, only a few hundred degrees Celsius. When ether burns, carbon dioxide and water are usually formed, but in cool flames the oxidation is only partial and the final products include acetic acid and acetaldehyde. The danger of cool flames is that they can transmit the risk of ignition to the anaesthetic breathing system.

In addition to cyclopropane and ether, there are other flammable anaesthetic agents such as ethyl chloride, ethylene and vinyl ether, but these are less commonly used. Some anaesthetic agents normally considered non-flammable will burn if mixed in the correct proportions with oxygen or nitrous oxide, and if a sufficiently high temperature occurs to ignite the mixture. Thus, the lower limits of flammability for halothane and trichloroethylene are higher than those likely to be used in normal anaesthetic practice, and also any sparks present are unlikely to be of sufficient energy to start the mixture burning.

Many anaesthetists now use only non-flammable anaesthetics, but this should not give rise to complacency because many other flammable agents may still be brought into the theatre area. Perhaps the commonest is ethyl alcohol or surgical spirit. A saturated vapour of alcohol in air at normal room temperature gives a 6% mixture.

As shown in Fig. 23.8, a 6% mixture is near the stoichiometric concentration in air and carries a high risk of burning; the risk being increased further if oxygen or nitrous oxide is present. Accidents have also occurred when liquid spirit on the patient's skin has ignited. The pale blue flames may not be readily visible and severe burns may occur before the accident is noticed. Fires have also occurred with alcohol when it has been used in nebulised form to sterilise ventilators.

In addition to alcohol, there are flammable substances that are not anaesthetic agents but which may give rise to fire or explosion risks. Thus methane and hydrogen may be present in the patient's gut and may be ignited by diathermy when the gut is opened. Ethyl chloride may be used as a local anaesthetic spray and ether to degrease the patient's skin. In addition to these flammable agents many other combustible materials such as paper, plastic drapes and dressings are found in theatre.

Pure oxygen under pressure results in exceptionally high fire and explosion risks. Hence, oil and grease must be avoided in any part of the compressed oxygen supplies, and special precautions are necessary in the areas of oxygen manifolds or liquid oxygen supplies.

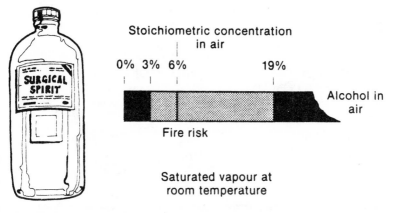

Figure 23.8 Flammability limits of ethyl alcohol in air.

ACTIVATION ENERGY AND IGNITION TEMPERATURE

In a mixture of a combustible agent and oxygen the molecules are normally kept apart by repulsive forces. For the mixture to burn, the molecules must be brought close enough to react together (Fig. 23.9). One obvious way of doing this is to increase the speed with which the molecules collide; this may be accomplished by increasing the temperature of the mixture locally, either by a spark or some other source of heat. The temperature must be raised to a sufficient degree over a sufficiently large volume and the energy required to bring this about is the activation energy.

The temperature to which a mixture must be raised for combustion to start is known as the ignition temperature. It varies with the constituents of the potentially explosive mixture and the proportions of the components. At the stoichiometric concentration of many explosive mixtures the ignition temperature is about 400 °C, but ether–oxygen mixtures can ignite at a temperature as low as 150 °C. This could occur if the mixture were brought into contact with a hot surface or naked flame—such as a spirit lamp used in ophthalmic surgery. Diathermy is also an obvious source of heat, and lasers (Chapter 24) too can ignite flammable materials.

Sparks may also provide the activation energy to initiate explosions.

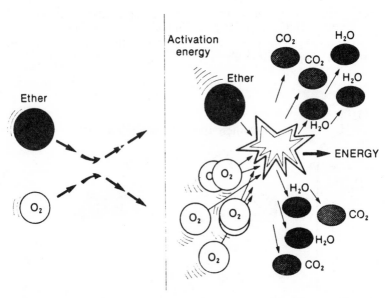

Figure 23.9   Activation energy increases molecular speed to allow the reaction to occur.

They are highly efficient in doing this, as, in contrast to a hot wire or naked flame, a spark concentrates the energy into a very small volume. Sparks with an energy as little as 1 microjoule can activate an ether–oxygen explosion when these gases are present in stoichiometric concentrations. One microjoule is a tiny amount of energy, being the amount needed to raise the temperature of 1 millilitre of water one-fifth of one-millionth of a degree Celsius.

Spark energy depends on the voltage of the source. A standard laryngoscope has a low voltage battery and no explosions have been reported when intubating patients using laryngoscopes in the presence of ether. In contrast, surgical diathermy uses high voltages. Even stray diathermy currents may cause sparks in the area of the patient's airway and anaesthetic connections, and consequently flammable agents must never be used in the presence of diathermy. In x-ray departments flammable agents are usually avoided because the possibility of sparks occurring in the x-ray apparatus cannot be eliminated. In this case full antistatic precautions, including an antistatic floor (see below), are not enforced because the cost is not justified in view of the fact that sources of ignition are present.

STATIC ELECTRICITY

Static electricity is the other important source of electric sparks. To prevent the build-up of static electricity the floors of operating theatres are electrically conductive to allow it to drain away. Too low a resistance is undesirable as it could increase electrocution risks, while too high a resistance would allow electrostatic charges to build up. To allow charges to be conducted to the floor, all apparatus and tables have conducting wheels or supports and staff wear antistatic rubber-soled shoes or clogs of the appropriate standard. Antistatic rubber contains carbon to make it electrically conductive and is also used for anaesthetic tubings. Materials with appropriate antistatic properties are also recommended for clothing and drapes—e.g. cotton is preferable to plastic. Endotracheal tubes, however, are not made of antistatic materials but are rendered conductive by the high humidity of the gases passing through them.

CHAPTER
24

Isotopes and Radiation

ATOMIC STRUCTURE AND ISOTOPES

An atom consists of a central nucleus containing protons that are positively charged and neutrons with no charge, and this nucleus is surrounded by orbiting electrons which have a negative charge. The type of atom, i.e. element, is determined by the number of protons (the atomic number); for example, hydrogen has 1 proton, carbon 6 and uranium 92. The neutrons contribute to the stability of the nucleus and, for a particular element with a given number of protons, different forms with different numbers of neutrons may exist. These different forms of the same element are called isotopes. An isotope of an element can be described by the total number of neutrons and protons which the nucleus contains, i.e. the mass number—examples of isotopes include hydrogen-1, hydrogen-2 (deuterium) and hydrogen-3 (tritium), carbon-12 and carbon-14, uranium-235 and uranium-238. Many isotopes of

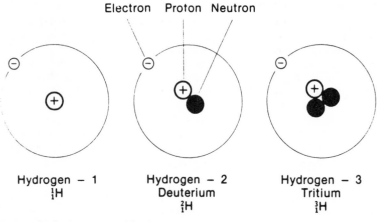

Figure 24.1   Isotopes of hydrogen.

elements occur naturally, but others may be produced artificially in a nuclear reactor.

Isotopes are often described by a special method and the three isotopes of hydrogen are used here as an example: ^1_1H, ^2_1H, and ^3_1H, the upper figure being the mass number and the lower the atomic number. The three isotopes of hydrogen are shown in Fig. 24.1.

The isotopes hydrogen-1 and deuterium are stable, but in the third, tritium, the nucleus is unstable and will not hold together indefinitely. Such an isotope is called a radioisotope, and is radioactive. The unstable nucleus of a radioisotope, also known as a radionuclide, eventually changes into another element in one of several ways. In the case of tritium, one of the neutrons in the nucleus changes into a proton and an electron; this electron is ejected from the atom (Fig. 24.2). The nucleus then contains two protons and one neutron, and the atom of tritium has changed into an atom of the stable element helium-3.

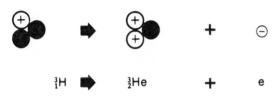

Figure 24.2 Radioactive decay of tritium. The tritium nucleus decays to a helium-3 nucleus and a beta particle.

MODES OF DECAY

The process of one element changing into another is known as radioactive decay and may occur in one of the following ways.

(1) An electron, known as a beta (β) particle, may be emitted (as in the example above). Although an electron is negatively charged, a positively charged beta particle known as a positron may be emitted by some nuclei.

(2) The nucleus may capture one of the electrons surrounding it.

(3) An alpha (α) particle may be emitted. An alpha particle is a combination of two protons and two neutrons (i.e. a helium-4 nucleus).

(4) The nucleus may undergo spontaneous fission into several fragments.

These four alternatives are not mutually exclusive and several types of decay may occur alternatively or in sequence.

After a nucleus has decayed by one of these processes it is usual for the nucleus of the new element formed to emit one or more gamma (γ) rays. A gamma ray is not a particle but an electromagnetic wave (Chapter 13). However, because it has a high frequency and short wavelength, it behaves in many respects like a particle.

If there is a delay between the decay of the nucleus and the emission of the associated gamma rays, the nucleus is said to be in a metastable state until the gamma rays have been emitted. The metastable state has its own rate of decay or half-life (Chapter 5) and the metastable isotope is identified by an 'm' after the mass number, e.g. technetium-99m.

The SI unit of radioactivity is the becquerel.

THE   BECQUEREL (Bq)

A given quantity of radioactive substance has an activity of 1 *becquerel* if one disintegration of a nucleus takes place on average every second.

Formerly, radioactivity was defined in terms of the curie (Ci), a curie being the quantity of any radioactive substance undergoing $3\cdot7 \times 10^{10}$ disintegrations per second.

DURATION OF RADIOACTIVITY, THE HALF-LIFE

Radioactive decay is a random process but the rate of decay at any moment will depend on the amount of radioactive isotope present, and consequently the process is exponential (Chapter 5). The rate of decay is measured by the half-life, i.e. the time required for half the radioactive atoms present to disintegrate. As an example, chromium-51 used in red-cell volume measurement (Chapter 3) has a half-life of $27\cdot8$ days. Thus, if there are initially 10^{12} atoms of chromium-51 in a sample of a chromium compound, there will be 5×10^{11} remaining after $27\cdot8$ days. After a further $27\cdot8$ days there will be $2\cdot5 \times 10^{11}$ atoms of the isotope remaining and so on.

CHARACTERISTICS OF EMITTED PARTICLES AND RADIATION

The alpha and beta particles and the gamma rays emitted by a radioactive isotope have a particular energy which is characteristic of the radioisotope concerned. The energy of electromagnetic waves is proportional to their frequency, and gamma rays and x-rays are also usually referred to in terms of their energy rather than their frequency. If the radioactive substance is within the body of the patient, all or part of this energy may be absorbed by the tissues. All the energy of alpha and beta particles is absorbed by the surrounding tissues causing damage, but in the case of gamma rays a proportion of the radiation escapes from the body and may be measured by a suitable detector. On the other hand, if a radioactive isotope emitting gamma radiation is outside the body, the gamma rays will penetrate the skin and be absorbed there and in the deeper tissues. A radioisotope can be sealed in a suitable container so that gamma rays only are emitted and any associated alpha and beta particles are absorbed in the container walls. Such sealed sources are used in the treatment of disease.

Alpha and beta particles and gamma radiation all cause damage to or death of cells. Actively dividing cells, such as tumour cells, are particularly vulnerable.

DETECTION OF RADIATION

Radiation may be detected using a scintillation counter, which is shown in Fig. 24.3. Gamma rays are absorbed in the sodium iodide crystal and their energy is converted into a flash of light. The intensity of the light is converted into an electrical pulse and amplified greatly by the photomultiplier tube.

The radioactivity in fluid samples containing isotopes which emit only beta radiation may be measured by mixing the sample with a liquid scintillator, which is a compound that will emit a flash of light when a molecule is struck by a beta particle. The flashes of light may then be detected and counted.

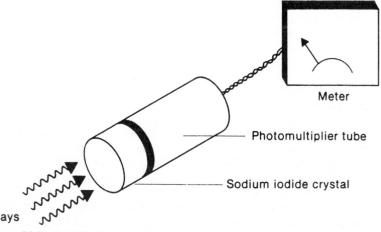

Meter

Photomultiplier tube

Sodium iodide crystal

γ rays

Figure 24.3   Scintillation counter.

THERAPEUTIC USES OF RADIOISOTOPES (RADIONUCLIDES)

In the treatment of tumours the radiation may be given by means of an external or internal source. The usual external source is a sealed capsule of cobalt-60 which has a half-life of around five years. In some cases an internal source of radiation is placed in the patient. Thus, a sealed yttrium-90 implant is used in the treatment of pituitary tumours which have caused acromegaly, and a sealed caesium-137 capsule is implanted in the uterus in treatment of some uterine tumours. Caesium-137 is also sometimes used with an 'after loading' technique. In this, cannulae are first placed in a tumour (e.g. of breast) and, once in position, caesium-137 needles are inserted down the cannulae to provide a high localised dose of radiation.

In addition to the therapeutic use of isotopes in the treatment of neoplasms, they are sometimes used to treat other conditions. Examples are the use of oral iodine-131 in the treatment of thyrotoxicosis or intravenous phosphorus-32 for polycythaemia rubra vera.

DIAGNOSTIC USES OF RADIOISOTOPES (RADIONUCLIDES)

Diagnostic techniques using radioactivity may be divided into two groups, imaging and non-imaging.

Imaging Techniques

Imaging techniques use the gamma rays emitted from an isotope and an image of the distribution of a radioactive compound within the body can be obtained by using a gamma camera. The gamma camera is similar to the scintillation counter; the gamma rays interact with a large crystal of sodium iodide causing the emission of flashes of light that can be detected electronically. From the positions of these flashes in the crystal a picture of the isotope distribution in an organ may be built up.

Technetium-99m is the radioactive isotope commonly used in imaging techniques because it has the advantage of being easily attached to various chemical compounds which can be injected into the patient. It also has a short half-life, thus reducing the radiation dose to the patient, and it produces gamma rays of an energy suitable for forming a good image. Thus, technetium-99m labelled albumin or red blood cells may be used for vascular imaging and for the assessment of cardiac function, while labelled micro-aggregates of human serum albumin may be used in a lung scan to assess pulmonary circulation. Technetium-99m pertechnetate is used in thyroid and brain scanning, and other technetium-labelled compounds are used in liver, spleen and bone scanning.

An alternative form of lung scan uses the radioactive gases xenon-133 or krypton-81m. If a small quantity of one of these gases is inhaled, an image of its distribution allows the assessment of regional ventilation. If the gas is injected in solution, it will be excreted into the alveoli to give a measure of regional differences in pulmonary perfusion.

A different type of imaging technique has been used in experimental animals to investigate the uptake of anaesthetic agents. The anaesthetic agent is labelled with a radioisotope such as carbon-14 and after an animal has inhaled the anaesthetic, it is killed and sectioned. The sections are placed in contact with photographic film and, after development, the film is found to be darkened in areas due to exposure to the radiation from the carbon-14. An estimate of the relative uptake by various organs can then be made from the relative exposure produced on the photograph.

Non-imaging Techniques

Non-imaging techniques using radioactivity include the use of chromium-51 labelled red cells or albumin for the measurement of red cell volume or plasma volume, respectively. A small known quantity of radioactive red cells or albumin is injected and after a suitable interval the dilution of the radioactive tracer in the patient's blood can be measured and from it the red cell or plasma volume can be calculated (Chapter 3).

The blood flow into or out of an organ can be measured if a radioisotope is used as an indicator. Suppose an organ contains a given quantity of a radioisotope at a particular time. The rate at which the radioactivity decreases in the organ depends on the blood flow and the volume perfused by this flow. The situation is analogous to the washing out of muddy bath water considered in Chapter 5, and a plot of radioactivity against time is a washout curve, from which the time-constant and thus the blood flow may be determined.

Fibrinogen labelled with radioactive iodine-125 may be used to detect deep venous thrombosis. The labelled fibrinogen is injected intravenously into the circulation. Should a clot begin to form, fibrinogen is locally concentrated within the clot. A proportion of this fibrinogen is the radioactive fibrinogen, which may be detected by means of the gamma rays it emits. A portable scintillation counter moved along the deep veins in the legs will show a locally high reading above a blood clot. In this test it is necessary to block the uptake of radioactive iodine by the thyroid by giving the patient a large dose of potassium iodide beforehand.

STABLE ISOTOPES

As stable isotopes are not radioactive they are difficult to isolate or detect and have few special uses in medicine. They can be identified by the mass spectrometer described in Chapter 20.

X-RAYS

X-rays, like gamma rays, are electromagnetic radiation but are produced in a different way (Fig. 24.4). They are produced when a beam of electrons is accelerated from a cathode to strike an anode, usually of tungsten. A divergent beam of x-rays is produced, and a lead shield with a window in it is used to restrict the size of the beam. X-rays are used for imaging purposes and in the treatment of disease.

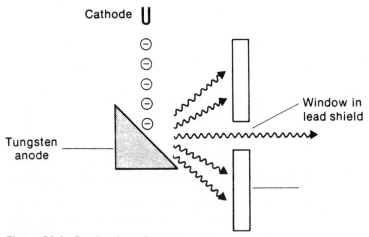

Figure 24.4   Production of x-rays.

RADIATION SAFETY

As alpha and beta particles, gamma radiation and x-rays may all cause tissue damage and chromosomal changes it is important that exposure to them is kept to a minimum. A person may be exposed to radiation from two sources; radioactive elements within the body, and radioactive elements and other sources of radiation outside the body.

The ingestion of radioactive materials may be avoided by safe handling procedures designed to ensure that such materials are not allowed to contaminate surfaces where they are being used. Typical precautions would include wearing gloves by people handling the materials, and arrangements to contain spillage of radioactive liquids.

Exposure to external radiation may be reduced by enclosing radioactive sources in containers made of material that absorbs radiation. Alpha particles travel only a few centimetres in air before their energy is expended and therefore the principal problem is one of containment to ensure that no alpha-emitting material is inadvertently inhaled or ingested. Beta particles can travel a few metres in air before they are absorbed; hence shielding a source with radiation-absorbing material is necessary. Because beta particles produce x-rays if de-celerated rapidly it is best to provide shielding with a material of comparatively low density, such as perspex, to avoid unnecessary conversion of the beta particle energy into x-rays. Gamma rays can travel large distances and shielding gamma ray sources is best achieved with dense materials such as lead. A thickness of 7 cm lead will absorb at least 90% of gamma rays of the type emitted by radioisotopes. Lead also absorbs x-rays, so it is incorporated into aprons worn by staff exposed to radiation. Local shielding may also be used to protect parts of patients outside the main x-ray beam when radiographs are being taken.

As actively dividing cells are particularly affected by radiation, the fetus is at special risk even before the mother may realise that she is pregnant. It is thus recommended that, except in emergency, women of reproductive capacity should be x-rayed only in the first 10 days of their menstrual cycle, i.e. before ovulation has occurred (the 10-day rule).

Apart from the effects of absorption, the intensity of any radiation is inversely proportional to the square of the distance from the source. Hence, the dose of radiation received may be reduced greatly by moving away from the source.

Regular checks for contamination of surfaces where radioisotopes are used should be carried out; a scintillation counter may be used. Where it is not easy to predict what a person's exposure to radiation will be or in circumstances where it is desired to check on the effectiveness of working practices, a person may wear a photographic film badge to monitor the total radiation dose received. The film badge contains a small piece of photographic film behind several different filters and permits estimation of the energy and dose of radiation received.

Hospitals have available to them the services of a radiation protection adviser, who should be consulted before radioactive compounds are used; in the UK a doctor requires a certificate of authorisation before he can administer a radioactive compound to patients. The symbol shown in Fig. 24.5A is displayed when there is a risk of exposure to ionising radiation.

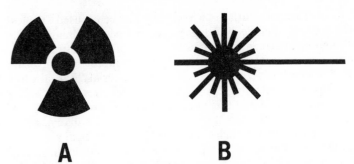

Figure 24.5 Warning symbols for ionising radiation (A) and for lasers (B).

NON-IONISING RADIATION

X-rays and gamma rays are known as ionising radiations because their energy is dissipated in tissues initially through the process of the removal of electrons from atoms. The resulting ions are very reactive chemically, so ionising radiation is particularly hazardous. Radiation from other parts of the electromagnetic spectrum dissipates its energy in tissues in other ways, and is therefore known as non-ionising radiation.

Non-ionising radiation comprises radiowaves, microwaves, infra-red radiation, visible light and ultra-violet radiation. There are also hazards associated with the use of non-ionising radiation, and so protection measures must be considered for all types.

A special risk of burns from infra-red or visible light radiation arises during the use of lasers. Lasers allow a beam of radiation to be focused onto a very small area, typically a spot of 0·1 to 2 mm diameter depending upon the requirements. They are used in ophthalmology to treat retinal detachment or vascular diseases of the retina, in dermatology to treat port wine stains, and in gynaecology to treat neoplasms, e.g. of the cervix. In otolaryngology, lasers may be used to treat lesions in the mouth or of the larynx; they have also been used to remove tumours obstructing the trachea or bronchus.

In view of the safety hazards they present the principle of the laser is considered in more detail.

PRINCIPLE OF THE LASER

As shown in Fig. 24.6 a gas laser is similar to a fluorescent light tube. In a laser, a high voltage is applied between two electrodes at the ends of a long narrow tube containing a gas such as argon or carbon dioxide. The current flow through the tube places the atoms of gas into a state in

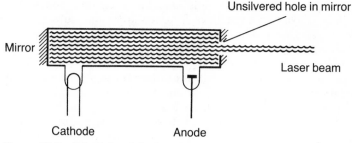

Figure 24.6 Principle of the laser.

which they possess more energy than normal; this is known as the excited state. Excited states have a certain fixed energy difference from the normal or ground state. Consequently, when an atom returns from an excited state to the ground state, the energy is often dissipated as light or radiation of a specific wavelength characteristic of the atom.

In a laser, the radiation may be in the visible or the invisible infra-red part of the spectrum and is reflected back and forth along the length of the tube by two mirrors which are accurately parallel to each other at the ends of the tube. The intense radiation causes the atoms to return to the ground state and emit radiation faster than would otherwise be the case. The radiation they emit is exactly in phase with the radiation already passing up and down the tube. The result is the conversion of the electrical energy into a beam of radiation which escapes from the tube through a small unsilvered area of one of the mirrors, forming an intense, narrow, parallel beam. Some lasers use a solid rod of ruby crystal or of neodymium yttrium aluminium garnet in which the atoms are excited by a discharge lamp beside the rod.

The radiation from a laser has a specific wavelength; for example, the argon laser gives visible light and is used in ophthalmology, while the carbon dioxide laser gives infra-red radiation and is more powerful and suited to surgical use when excision of tissue is needed. In some lasers, radiation is conducted along optical fibres to the point where treatment is needed. However, these fibres are opaque to the infra-red radiation from the carbon dioxide laser, so its beam must be guided by a system of mirrors.

OPTICAL FIBRES

As mentioned above, optical fibres can be used with lasers to direct the beam, but they are also used in the design of endoscopes and bronchoscopes to permit the physician to see round corners.

The principle of their action is as follows. The direction in which light is travelling is bent (refracted) as it passes from one substance to another. If the angle between the direction in which the light is travelling and the boundary between the two media is sufficiently acute, the light may be totally reflected (Fig. 24.7). Consequently, if light passes into one end of a fibre of glass or other transparent material, it can pass along the fibre by being continually reflected from the glass/air boundary. Fibreoptic instruments such as endoscopes use bundles of flexible fibres to transmit the light easily round corners. Optical fibres may be used to direct the high intensity output from lasers to the site where it is needed.

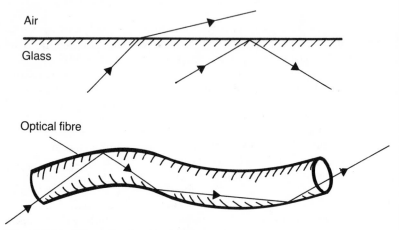

Figure 24.7   Principle of optical fibres.

LASER SAFETY

The effect of lasers on the retina of the eye is used therapeutically as mentioned above. However, the eye is also the organ most at risk when lasers are used, because the beam may be focused by the lens onto the retina and cause irreversible damage in a very short time. Because the carbon dioxide laser beam is absorbed within 200 µm of the tissue this laser is a hazard to the cornea rather than the retina. The laser beam can be accidentally reflected into the eye from a shiny surface such as an operating instrument, so matt surfaces are preferred and well-fitting protective goggles which absorb radiation of the wavelength at which the laser operates should always be worn. Skin protection of personnel is not usually necessary but the patient's tissues adjacent to the lesion to be treated should be protected by water soaked pads.

A low power or secondary light beam is used to aim the laser and the operator should only fire the beam of laser radiation when it is correctly aimed. All staff present should be warned when the laser is about to be fired and a special warning sign should be conspicuously displayed outside the door of the operating area (Fig. 24.5B).

The other danger is that a high power laser beam is capable of igniting flammable materials such as rubber or plastic. Tubes carrying oxygen or anaesthetic gases may be damaged by the laser beam and should be protected by wrapping them with metal tape or damp muslin. Alternatively, special non-combustible laser resistant tubes have been produced. Local rules for the operation of each laser should be drawn up after discussion with a laser protection adviser.

25

Presentation and Handling of Data

LINEAR AND NON-LINEAR SCALES

The scales on many instrument dials are linear (Fig. 25.1A), but sometimes the scales are non-linear, being expanded at lower readings to give greater sensitivity. An example is the scale on the oxygen flowmeter of many anaesthetic machines (Fig. 25.1B) in which the conical bore of the interior of the flowmeter is moulded so that low flows of 100 to 500 ml per minute register on an expanded scale. This makes adjustment easy both for the low flows used in a closed circuit and for the higher flows used in the Magill system.

A special form of non-linear scale is the logarithmic type (Fig. 25.1C and D) in which each linear increment on the scale corresponds to a proportionate increase in the amount measured. This scale also has the advantage of allowing greater accuracy when measuring small quantities, while preserving the ability to cover a large range.

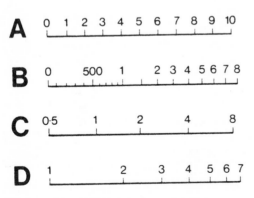

Figure 25.1 Different types of scale.

LOGARITHMIC SCALES

In addition to their use on analogue displays, logarithmic scales are used in digital format to achieve the same advantages of a convenient method of covering a wide range of a quantity. A digital logarithmic scale with 10 as the base can be identified by the exponent or power of 10 as follows:

10^{-3}	10^{-2}	10^{-1}	10^0	10^1	10^2	10^3
0·001	0·01	0·1	1	10	100	1000

One example of the use of a logarithmic scale to the base 10 is pH (Chapter 19), and the use of exponents is considered in Chapter 5. The use of an exponent to the base 10 allows large and small quantities to be presented more compactly and more conveniently than using a large number of digits. Thus, a quantity of 0·0000000251 can be written as $2·51 \times 10^{-8}$, and 573 000 000 can be expressed as $5·73 \times 10^8$.

Many calculators switch automatically into a display indicating the exponent to the base 10 in place of the decimal point when the quantity is too large or too small for the display.

In the SI system it is recommended that the units are expressed in terms of multiples or submultiples of a thousand 10^3, e.g. kilo (k) for a thousand times and milli (m) for a thousandth. The names and symbols of most of the common multiples in use are given in the appendix. The multiple is usually chosen so that the numerical value lies within the range 0·1 to 1000; for example:

$$137\,000\,000 \text{ Pa} = 137 \text{ MPa}$$

$$0·002 \text{ g} = 2 \text{ mg}$$

Logarithmic scales are also used in graphs and an example of the use of a semi-logarithmic plot, in which one axis is logarithmic and the other linear, is considered in Chapter 5. A general property of all logarithmic scales is that they cannot extend to zero.

GRAPHICAL DISPLAYS OF DATA

In circumstances such as the inflation of the lungs by a ventilator there are several quantities or variables that change in value as inflation proceeds. A list of these variables might include gas flow, airway pressure, alveolar pressure, lung volume and time. Often a change in one variable will affect the value of one or more of the other variables—

i.e. the variables are not independent, but the value of one variable is related to or is a function of the value of the others.

A graph is a convenient way of showing how one variable depends on another. It has two axes, each of which is assigned to a variable, and on each axis a scale is marked showing the values taken by that variable. The horizontal axis is called the abscissa or the x-axis and the vertical axis is called the ordinate or the y-axis.

Figure 25.2A illustrates a graph of the way in which inspired volume changes with airway pressure during slow inflation of the lungs in an anaesthetised relaxed patient on a ventilator. In this example it is assumed that the rate of inflation is so slow that the pressure difference due to the airway resistance can be ignored. Electrical signals of the two variables, volume and pressure, could be obtained by an electronic Wright respirometer and by a pressure transducer, and the graph could be drawn automatically by an x–y recorder. This type of recorder accepts two electrical inputs for the two variables and plots them simultaneously on a graph.

In this example the volume increases linearly with increasing pressure and the formula which describes the relationship of volume to pressure is:

$$V_I = k_1 P$$

where k_1 is a constant equal to the slope of the graph which, in this case, is V_I/P, the compliance of lungs and chest wall.

If the total volume of air in the lung is plotted, then the graph is modified as shown in Fig. 25.2B and a volume k_2 equal to the functional residual capacity of the lung is added to the inspired volume at each point on the graph as shown.

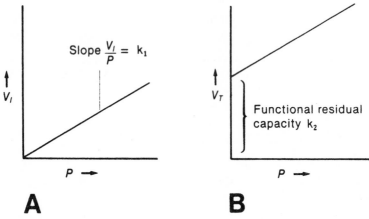

Figure 25.2   Linear graphs.

The formula for the graph then becomes:

$$V_T = k_1 P + k_2$$

A straight line graph can always be described by an equation of this type. V_T and P are the measured variables in this case, V_T being a function of P; k_1 is the slope or gradient of the graph (in this case the compliance), k_2 is the intercept on the ordinate or y-axis (in this case the functional residual capacity).

Although k_1 and k_2 are constants, they have different values in different circumstances and in different patients. Such constants are known as parameters. The term parameter is sometimes misused to refer to the variables being measured, but this is best avoided.

NON-LINEAR GRAPHS

For practical purposes in turbulent flow (Chapter 2) the following formula describes the relationship between pressure (P) and flow ($\dot{V}$, rate of volume change).

$$P = k \dot{V}^2$$

Figure 25.3A is a graph of this equation and is known as a parabola. A parabola often traces a U shape as shown in Fig. 25.3B and may be

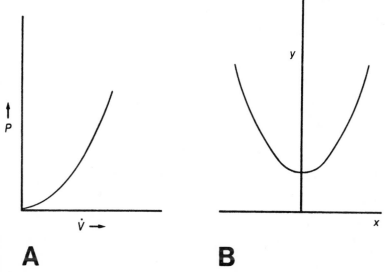

Figure 25.3 Parabolic curves.

described in terms of parameters in a similar way to the straight line graph. The general formula for a parabola is:

$$y = a + bx + cx^2$$

where y and x are the variables and a, b and c are the parameters.

Other examples of non-linear graphs are the rectangular hyperbola where $y = k/x$, as exemplified by Boyle's law (Chapter 4), natural exponential functions, e.g. $y = e^{-kx}$ (Chapter 5), and sine waves where $y = k$ sine x (Chapter 13).

SLOPE OF A NON-LINEAR GRAPH

Figure 25.4 shows a graph of the volume expired against time for a rapid expiration. Such a tracing could be obtained with a spirometer (Chapter 3). The slope of the graph at any point represents the expiratory flow at that point and the steepest slope, as shown, is the peak expiratory flow. The slope is obtained by drawing a tangent to the curve.

The slope of any graph of volume against time equals the flow, but it is time-consuming and relatively inaccurate to draw tangents to deduce slope and thus flow. Provided that the equation of the graph is known, a mathematical technique called differentiation may be used to derive the gradient or slope of the graph at any point. Monitors sometimes include electronic processing so that variables can be differentiated if required, e.g. to produce flow from volume change.

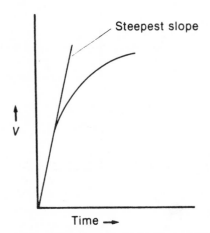

Figure 25.4   Slope of non-linear graph.

AREA BENEATH A GRAPH

Consider the plot of flow, $\dot{V}$, against time illustrated in Fig. 25.5. This could be produced during an expiration, with a pneumotachograph being used to measure the flow. The area under such a graph equals the product of average flow and time, and so is the volume expired. The shaded area on the graph may be measured by counting squares or by using an instrument known as a planimeter, but the area can also often be deduced mathematically provided the formula for the graph is known. The mathematical process used is known as integration.

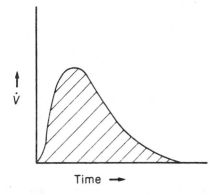

Figure 25.5 Area beneath a graph of flow ($\dot{V}$) against time represents volume.

Integration may also be used in calculating the area beneath a dye dilution curve in measurements of cardiac output (Chapter 5), and in measurements of the area of a volume–pressure tracing when calculating the work of breathing (Chapter 8). As in the case of differentiation, facilities to integrate variables electronically may be incorporated in monitoring apparatus.

PIE DIAGRAM

Graphs are not the only way in which data can be presented diagrammatically, and a pie diagram is a useful technique to illustrate how a certain quantity is divided up. Figure 25.6 illustrates the distribution of the blood flow in the body and gives an immediate impression of the relative importance of the individual organs. The same data could be provided as a table of percentages of blood flow to

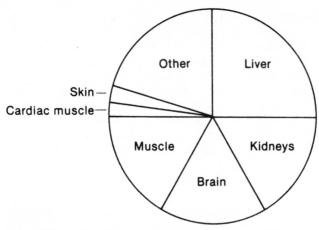

Figure 25.6 Pie diagram illustrating blood flow distribution in the body.

each organ. However, a pie diagram gives a clearer visual impression of blood flow distribution.

HISTOGRAM

In a histogram, the height (or area) of bands or columns is used to illustrate the relative sizes of quantities. In Fig. 25.7 the bands are used to illustrate the relative mean haemoglobin concentrations of adult men

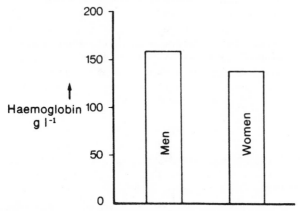

Figure 25.7 Histogram illustrating the haemoglobin concentrations of men and women.

and women and could, for example, represent the mean results for several hundred estimations in a laboratory. The histogram gives good visual impact and is particularly useful for illustrations used in teaching. However, in the unmodified form shown the histogram gives no indication of the range of results, and without additional data does not prove that the mean haemoglobin of the men is significantly greater than that of the women.

An alternative form of histogram (Fig. 25.8) can be used to illustrate the frequency at which different values of a variable occur. Suppose, for

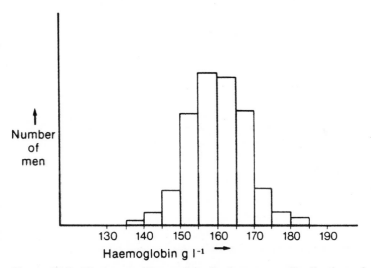

Figure 25.8 Histogram illustrating the frequency distribution of haemoglobin concentration of a large population of men.

example, that instead of plotting the haemoglobin value as the height of the histogram, the number of men with a given haemoglobin concentration in each 5 g litre^{-1} range is plotted as the height of each band of the histogram. For a survey of 100 fit adult men the histogram could show a spread of results as shown in Fig. 25.8; and it is seen that the histogram traces a bell-shape.

NORMAL DISTRIBUTION CURVE

If haemoglobin values were taken from an even larger population of several thousand and plotted at small increments, then an even smoother bell-shape would result. In the case of many physiological measurements the shape of this distribution approaches that of a mathematical curve known as the normal distribution curve (or Gaussian curve) shown in Fig. 25.9. Such a curve can be described in terms of its mean value, which is also its most frequent result, and by its breadth.

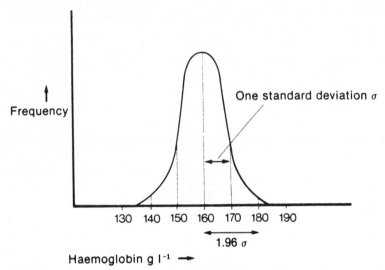

Figure 25.9   Normal distribution curve for haemoglobin concentration in men.

The mean of any set of results is formed by taking the sum of all the individual values and dividing by the number of values and is often expressed as:

$$\text{Mean, } \bar{x} = \frac{\Sigma\, x_i}{n}$$

where Σ means sum of, x_i is an individual result and n the number of results.

The breadth of the curve is a measure of the variability of the quantity concerned and a statistical term, standard deviation—often given the symbol σ (sigma)—is used as a measure of it.

The standard deviation of any set of results can be found from the formula:

$$\text{Standard deviation, } \sigma = \sqrt{\frac{\Sigma\,(x_1 - \bar{x})^2}{n - 1}}$$

but most modern calculators can produce σ automatically from a set of results.

Suppose this distribution of the haemoglobins of a large number of men was plotted and was found to be a normal distribution with a mean of 160 g litre^{-1} and a standard deviation of 10 g litre^{-1}. It follows from the shape of the normal distribution that 68% of the men have a haemoglobin between 150 g litre^{-1} and 170 g litre^{-1}, that is within one standard deviation from the mean, whereas 95·44% of the men have a haemoglobin between 140 g litre^{-1} and 180 g litre^{-1}. A more commonly used range is 1·96 standard deviations on each side of the mean which includes 95% of the results. This range is known as the 95% confidence range, the limits being known as the 95% confidence limits.

If a result occurs which has a value outside the 95% confidence limits, it belongs to a 5% minority in that population. Put another way, the probability of finding a haemoglobin outside the 95% confidence limits is 5%, which is written $P < 0.05$. The result can then be said to be significant at the 5% level. If $P < 0.01$ the result is significant at the 1% level.

The same concepts and calculation of standard deviation may be applied to other measurements such as temperature, blood pressure, electrolytes and carbon dioxide tension.

SOURCES OF VARIABILITY

The variability in a set of results arises from several sources: the individuals from whom the measurements are taken; the sampling or recording techniques used; and the measuring apparatus. A distribution curve could be drawn for each of these sources to indicate the variability of each (Fig. 25.10), but it is found that the relationship

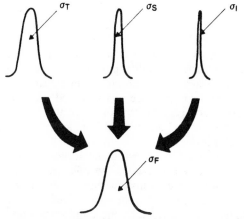

Figure 25.10   Analysis of variance. $\sigma_F^2 = \sigma_T^2 + \sigma_S^2 + \sigma_I^2$.

between the total variability in the measured values and that of the component sources depends on the square of the standard deviation, which is known as the variance (σ^2).

$$\sigma_F^2 = \sigma_T^2 + \sigma_S^2 + \sigma_I^2$$

where σ_F^2 = variance of final measurement
σ_T^2 = variance of total population under study
σ_S^2 = variance of measuring or sampling technique
σ_I^2 = variance of measuring instrument

The analysis of the component parts of variability in this way is known as an 'analysis of variance' and, in practice, the sources of variation may be more than the three given in this simple example.

THE NORMAL RANGE

Many laboratories provide normal ranges of values for various tests. Usually this consists of 95% of the most typical results of a survey of 'normal' individuals and may be expressed as a mean ± 1.96 standard deviations when the distribution approximates to that of a normal distribution. Such a range is known as the 95% confidence interval for the method.

The range includes variability arising from the patients and that from the sampling technique, the time of day at which the sample is drawn and, on some occasions, variation from monthly and annual cycles. The normal range should also include variability from the laboratory's instrument and its use, although this is usually a minor component of total variability for the most common tests.

INSTRUMENT RELIABILITY AND QUALITY CONTROL

Although instrument manufacturers provide accuracy ranges for their instruments, larger laboratories include quality control measures to ensure that the precision and accuracy of measurement is maintained in practice. Replicate quality control samples are incorporated into the routine work load and the results compared with those expected, taking into account the precision of the method as assessed from an extensive series of measurements on such samples.

A distinction is made between the terms precision and accuracy when assessing laboratory analysis. Precision is used when referring to the reproducibility of replicate results on a sample within a laboratory. Accuracy is a measure of the difference between the mean of such results and the true value of this material as obtained by the standard method of analysis.

In addition to the inherent variability in the processes of sampling and measuring a variable, it should not be forgotten that occasional gross blunders can occur, examples being mislabelling of specimens, a misplaced decimal point, or the misuse of reagents or apparatus. A good quality control scheme will check all facets of performance from patient to report, and not simply the analytical processes.

RELIABILITY OF THE MEAN OF A SET OF RESULTS

The mean of a set of results can be expected to show less variation than an individual result, and consequently the standard deviation of the mean is used in place of the standard deviation as an indication of the variability of the mean. The standard deviation of the mean is known as the standard error of the mean and is calculated as follows:

$$\text{Standard error of the mean, s.e.} = \frac{\sigma}{\sqrt{n}}$$

Figure 25.11 illustrates the use of the standard deviation and the standard error. In Fig. 25.11A the bar at the top of the histogram is used to indicate the variability of the haemoglobin results for men, so the standard deviation is used. In Fig. 25.11B the mean haemoglobin of 20 men is being compared to that of 20 women, so bars indicating the standard error of the mean are more appropriate. Since the ranges indicated by the standard error do not overlap but are substantially different from each other, it is unlikely that the difference is due to chance.

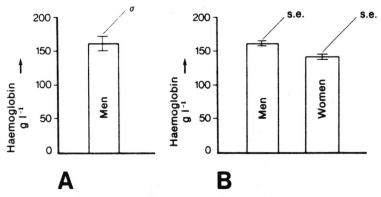

Figure 25.11(A) Histogram illustrating use of standard deviation (σ). (B) Histogram illustrating use of standard error (s.e.).

COMPARISON OF TWO SETS OF DATA

The Student's *t*-test may be used to compare two sets of normally distributed samples when the standard errors and means of the two sets are known. The test can be used in either of two forms, depending on whether two separate series of results are being compared (unpaired *t*-test) or whether results are in the same group of patients (paired *t*-test). An example of the latter could be a comparison of results before and after treatment.

Modern calculators allow direct calculation of *t* for unpaired or paired *t*-tests. The probability that the result is due to chance depends on both the value of *t* and the number of results and, consequently, tables must be consulted after the value of *t* is known.

The *t*-test is commonly used in clinical trials when the measurements have a normal, or near normal, distribution.

SKEWED DISTRIBUTION

Not all measurements show a normal distribution. Figure 25.12A shows a lopsided or skewed curve. Figure 25.12B represents haemoglobin results in a series of women patients in whom anaemia is relatively more frequent than high haemoglobin. It can be seen in each diagram that the most frequently occurring measurement, known as the mode, is not the same as the mean.

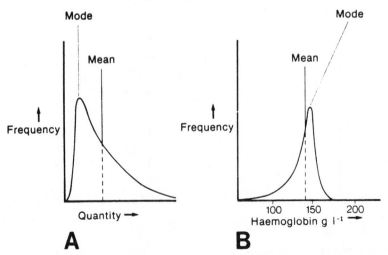

A   **B**

Figure 25.12 **Skewed distribution.**

If results have a skewed distribution, then comparison of two such sets by means of a Student t-test is inappropriate. There are other statistical techniques which can be used when comparing results that are not distributed normally. These alternative statistical techniques are called non-parametric because they do not assume anything about the parameters of the distributions being compared.

NON-PARAMETRIC TESTS

An example of a non-parametric test is the Wilcoxon rank sum test. It is used to assess whether there is any significant difference between two sets of measurements. Consider the example in the table below which gives the sleeping time in minutes in 20 patients randomly given different drugs, so that ten are given drug A and ten drug B.

Sleeping Time (minutes) Drug A	Rank	Sleeping Time (minutes) Drug B	Rank
100	5·5	92	9·5
83	13	78	16
126	1	63	19
65	18	59	20
92	9·5	89	11
81	14	101	4
111	2	76	17
100	5·5	95	7
79	15	94	8
103	3	87	12
940	86·5	834	123·5
Mean $\bar{x}_1$ 94·0		Mean $\bar{x}_2$ 83·4	

The two sets of measurements are ranked, the highest from either set being given the value 1, the next 2 and so on. If two values are identical the rank figure is averaged as shown. The sum of the ranks for the two sets of measurements can then be calculated. Consultation of tables then shows that the ranking figures 86·5 and 123·5 do not reach significance at the 5% level for the ten patients concerned in each group.

This example included data from measurements in two different groups of patients. In some experimental trials two sets of data may be obtained from the same patient. These are known as paired values. The statistical significance of the difference between paired values may be assessed using similar ranking techniques and examples include the Wilcoxon paired test and Spearman's rank test, but full details of these are beyond the scope of this book.

Another test is the chi-squared (χ^2) test in which the frequency of observed results is compared with the expected frequency. This test may be used when assessing the effectiveness of drugs in producing a particular result. If the expected result is regarded as no change, the probability of the obtained result being significant can be assessed. This is often called the null hypothesis.

SCATTER DIAGRAMS

When individual results are plotted as points on a graph they are usually scattered, although an underlying trend may be apparent. A scatter diagram (Fig. 25.13) thus gives an impression of how the results are distributed. The example shows results taken before and after treatment and it is helpful to include a line on the scattergram to indicate the line of identical performance where $x = y$ as shown. In the example in Fig. 25.13A, all the results are below the line of identical performance and so it is obvious that, despite the variability of the results, the post-treatment measurements were always lower than the pretreatment values. In Fig. 25.13B all the results are closely distributed around the line of identical performance and it is apparent that the treatment had no effect.

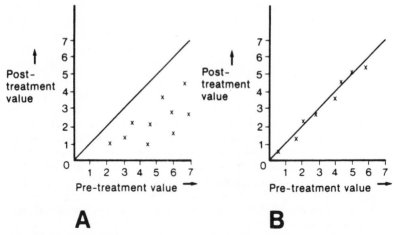

Figure 25.13 Scatter diagrams illustrating the use of the line of identical performance.

REGRESSION AND CORRELATION

The line providing the best fit to the points in a scatter diagram is known as a regression line; a measure of how well the points fit the line is the correlation coefficient. The correlation coefficient lies between 0 and 1, a value of 1 indicating perfect agreement between the points and the regression line and a value of 0 indicating no agreement between the points and the line.

A regression line may be linear or curved and it is misleading to fit a linear regression line to a set of points if it is obvious that the points indicate a possible non-linear relationship or if the scatter is so great that it precludes a decision on whether the relationship is linear or not.

USE OF STATISTICS

Any statistical test is useful only in the appropriate circumstances in a correctly designed experimental trial, and prior statistical advice should be obtained. It is too late once the trial has been completed to rectify the inadequacy of the design by the use of statistics.

COMPUTERS

Digital computers can carry out complex mathematical calculations at very high speed and hence permit rapid processing of large amounts of information which would otherwise be impractical to analyse. These calculations are carried out according to a programme, which is a set of specific instructions that the computer uses to produce the required results. The computer cannot operate without this programme which must either be written by the user or acquired from some other source.

As shown in Fig. 25.14, a computer contains a central processing unit (c.p.u.), which carries out the instructions in the programme, and a memory section which stores the programme—the data which the computer uses for its calculations and the results. The memory in a computer is of two sorts. The contents of random access memory (r.a.m.) may be changed as necessary as the computer operates, but the contents of such memory are lost when the computer is switched off. This is known as volatile memory. Read only memory (r.o.m.) retains its contents when the power supply is removed and is known as non-volatile memory.

Several peripheral items may be attached to the computer, including devices to store and read programmes and data in the form of magnetic discs and tapes, a keyboard to enter instructions into the computer and a visual display unit (v.d.u.) or printer to display the results. When

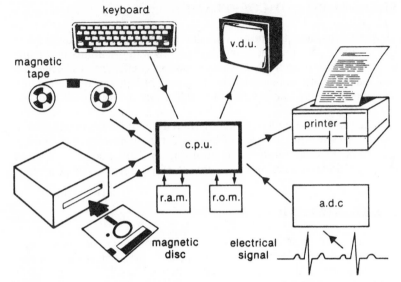

Figure 25.14   Diagram illustrating the digital computer and possible peripheral items which may be attached to it. Visual display unit = v.d.u.; central processing unit = c.p.u.; random access memory = r.a.m.; read only memory = r.o.m.; and analogue to digital converter = a.d.c.

calculations are carried out from stored data some time after this has been collected, the computer is used 'off-line'. As an alternative, a varying electrical signal may be converted into a digital form so that calculations may be carried out immediately. This is known as 'on-line' use and uses an analogue to digital converter (a.d.c.).

Although the input and output of numbers to a computer may be in the decimal system of counting, the computer itself uses the binary system (i.e. 2 is used as the base for counting rather than 10). The table below shows the representation of the numbers 0–10 in the decimal system and the binary system.

Decimal system	0	1	2	3	4	5	6	7	8	9	10
Binary system		0	1	10	11	100	101	110	111	1000	1001	1010

The usefulness of the binary system of counting is that it uses only the digits 0 and 1. These two digits may be represented in an electrical circuit by the absence or presence of a voltage, and a computer uses such a representation to perform the necessary arithmetic.

The set of binary instructions used by the central processing unit are known as machine code, but since it is difficult and time-consuming to write programmes in machine code several more easily understood sets of instructions, called languages, are used to write programmes. Examples of such high level languages are Basic, Fortran and Pascal. A programme that has been written in a high level language must be converted into machine code by a special programme in the computer before the central processing unit can carry out the instructions.

Analogue computers work on a different principle from digital computers and are less frequently used. In an analogue computer values are represented not by digital numbers but by the magnitude of electrical potentials.

Appendix

UNITS OF MEASUREMENT

The 'Système International d'unités' or SI system was introduced in 1960. As this is the system generally used in this book a resumé is provided.

Base SI Units

There are seven base units in the SI system from which all other units are derived. The base units are as follows with the symbol for the unit given in brackets.

(1) *Length* metre (m)
The metre is the distance light travels in a specified fraction of a second.

(2) *Mass* kilogram (kg)
The kilogram is based on the mass of a 1-kilogram prototype held at Sèvres near Paris.

(3) *Time* second (s)
The second is based on the frequency of radiation emitted from caesium-133.

(4) *Current* ampere (A)
The ampere is the current which produces a force of 2×10^{-7} newtons per metre between two conductors 1 metre apart in a vacuum.

(5) *Temperature* kelvin (K)
The kelvin is $1/273 \cdot 16$ of the thermodynamic temperature of the triple point of water. (The triple point is the point at which water vapour, ice and liquid water are in equilibrium.)

(6) *Luminous Intensity* candela (cd)
The candela is based on the luminous intensity of a perfect radiating black-body at a specified high temperature.

(7) *Amount of Substance* mole (mol)
The mole is the amount of substance which contains as many elementary particles as there are atoms in $0 \cdot 012$ kg of carbon-12.

Derived Units

From these seven base units numerous derived units may be obtained and examples of some of the more important are as follows.

Temperature degree Celsius (°C)
The size of the degree Celsius is identical to the kelvin and the relationship is:

$$°C = K - 273.15$$

Force newton (N)
A force of 1 newton will give a mass of 1 kilogram an acceleration of 1 metre per second per second.

$$1 N = 1 kg\, m\, s^{-2}$$

Pressure
    (i)—pascal (Pa)
A pascal is the pressure of 1 newton per square metre.

$$1 Pa = 1 N\, m^{-2}$$

    (ii)—bar (bar)
A bar is 100 kilopascals. It is not an SI unit, but has been retained for general use.

Energy joule (J)
A joule is the energy expended when the point of application of a force of 1 newton moves 1 metre in the direction of the force.

$$1 J = 1 N\, m$$

Power watt (W)
Power is the rate of energy expenditure. One watt is 1 joule per second.

$$1 W = 1 J\, s^{-1}$$

Frequency hertz (Hz)
One hertz is a frequency of 1 cycle per second.

Volume   (i) cubic metre (m^3)
             (ii) litre (litre)
The official SI unit of volume is the cubic metre (m^3), but the litre, one-thousandth of a cubic metre, is a permitted unit for general use.

$$1\text{ litre} = 10^{-3}\, m^3$$

Electrical Units

The main units used in this book are:

Potential volt (V)
One volt is the difference of electrical potential between two points of a

conductor, carrying a constant current of 1 ampere, when the power dissipated between these points is 1 watt.

$$V = W A^{-1}$$

Resistance ohm (Ω)
If a potential of 1 volt is applied across a conductor and produces a current flow of 1 ampere, then the resistance of this conductor is 1 ohm.

$$\Omega = V A^{-1}$$

Charge coulomb (C)
One coulomb is the quantity of electricity transported in 1 second by a current of 1 ampere.

$$C = A s$$

Capacitance farad (F)
A capacitor has one farad of capacitance if a potential difference of 1 volt is present across its plates, when a charge of 1 coulomb is held by them.

$$F = C V^{-1}$$

Units not in the SI System
Several unofficial units are still in common use.

Pressure (i) millimetre mercury (mmHg)
It should be noted that 1 kilopascal is equivalent to approximately 7·5 mmHg and 1 bar is therefore 750 mmHg.
(ii) centimetre water (cmH_2O)
One centimetre water pressure is approximately 98 Pa. (1 kPa $\simeq$ 10·2 cmH_2O.)
(iii) standard atmosphere (atm)

A standard or normal atmosphere is the pressure of 101·325 kPa. A bar is therefore about the same as an atmosphere.

Energy calorie (cal)
A calorie is 4·18 J. Calorie spelt with a capital 'C' may represent 1000 calories when it is used in dietary nutrition.

Force kilogram weight
The kilogram weight, sometimes called the kilopond, is the force of gravity on a mass of 1 kilogram. Gravity gives an acceleration of 9·81 m s^{-2}, consequently:

$$1 \text{ kilogram weight} = 9·81 \text{ N}$$

Multiples of Units
In the SI system the preferred multiples of the units are in terms of a

thousand and the commoner multiples and their symbols (in brackets) are as follows:

pico 10^{-12} (p)
nano 10^{-9} (n)
micro 10^{-6} (μ)
milli 10^{-3} (m)
kilo 10^3 (k)
mega 10^6 (M)
giga 10^9 (G)

Standard Temperature and Pressure (s.t.p.)
When expressed in SI units, s.t.p. is 273·15 K (0 °C) and 101·325 kPa.

Index